AF835594

HEAT EXCHANGERS– ADVANCED FEATURES AND APPLICATIONS

Edited by **S M Sohel Murshed**
and **Manuel Matos Lopes**

Heat Exchangers– Advanced Features and Applications
http://dx.doi.org/10.5772/68064
Edited by S M Sohel Murshed and Manuel Matos Lopes

Contributors
Antonio Carozza, L. Berrin Erbay, Mehmet Mete Öztürk, Bahadır Doğan, Kah Hou Teng, Salim Newaz Kazi, Abu Bakar Mahat, Bee Teng Chew, Ahmed Al-Shamma'A, Andy Shaw, Sinisa Sadek, Davor Grgić, Hafiz Muhammad Ali, Jong-Taek Oh, Chien Ba Nguyen, Yu Ito, S M Sohel Murshed, Manuel Manuel Luis Matos Lopes

CBS, Edition **2018**

Published by InTech
Janeza Trdine 9, 51000 Rijeka, Croatia

The moral rights of the editor(s) and the author(s) have been asserted.
All rights to the book as a whole are reserved by InTech. The book as a whole (compilation) cannot be reproduced, distributed or used for commercial or non-commercial purposes without InTech's written permission.
Enquiries concerning the use of the book should be directed to InTech's rights and permissions department (permissions@intechopen.com).
Violations are liable to prosecution under the governing Copyright Law.
© The Editor(s) and the Author(s) 2017

Individual chapters of this publication are distributed under the terms of the Creative Commons Attribution 3.0 Unported License which permits commercial use, distribution and reproduction of the individual chapters, provided the original author(s) and source publication are appropriately acknowledged. More details and guidelines concerning content reuse and adaptation can be found at http://www.intechopen.com/copyright-policy.html.

Notice
Statements and opinions expressed in the chapters are these of the individual contributors and not necessarily those of the editors or publisher. No responsibility is accepted for the accuracy of information contained in the published chapters. The publisher assumes no responsibility for any damage or injury to persons or property arising out of the use of any materials, instructions, methods or ideas contained in the book.

Publishing Process Manager Martina Usljebrka

Technical Editor SPi Global

Cover Designer InTech Design Team

Additional hard copies can be obtained from orders@intechopen.com

Heat Exchangers– Advanced Features and Applications , Edited by S M Sohel Murshed and Manuel Matos Lopes
p. cm.
Print ISBN 978-953-51-3091-8
Online ISBN 978-953-51-3092-5

PUBLISHED BY

INTECH

open science | open minds

World's largest Science, Technology & Medicine Open Access book publisher

2,950+
OPEN ACCESS BOOKS

100,000+
INTERNATIONAL AUTHORS AND EDITORS

96+ MILLION
DOWNLOADS

BOOKS
DELIVERED TO
151 COUNTRIES

AUTHORS AMONG
TOP 1%
MOST CITED SCIENTISTS

12.2%
AUTHORS AND EDITORS FROM TOP 500 UNIVERSITIES

CLARIVATE ANALYTICS
BOOK CITATION INDEX
INDEXED

WEB OF SCIENCE™

Selection of our books indexed in the Book Citation Index in Web of Science™ Core Collection (BKCI)

Interested in publishing with us?
Contact book.department@intechopen.com

Numbers displayed above are based on data collected at the time of publication, for latest information visit www.intechopen.com

Contents

Preface

This book presents contributions from renowned experts addressing research and development related to the major advanced features and fields of application of heat exchangers. It consists of nine chapters. Except the introductory chapter, which brings in perspective and overviews of the content of the book, the first three chapters start with advanced design and geometry features related to different kinds of heat exchangers; the next chapters deal with the use of compressible and supercritical thermal fluids followed by two chapters presenting two special applications and ending with advances in treating the problems of fouling and corrosion in heat exchangers.

This book starts with an introductory chapter giving a perspective of the importance of the heat exchanger technology and also highlighting all the contributions of this book, providing some complementary information.

The second chapter covers the two-phase flow pressure drop and heat transfer of a refrigerant during boiling in macroscale and miniscale channel-based heat exchangers.

The third chapter presents the study of the performance of compact heat exchangers with offset strip fin using different offset strip geometries.

The same authors follow in the fourth chapter with the description of the basic physical features and the analysis of the performance of the heat exchangers with louvered fins.

The fifth chapter reports the current state of knowledge of free-convection condensation heat transfer on geometrically enhanced tubes. It reports extensive comparative experimental work performed on integral and enhanced pin-fin tubes demonstrating that geometry is not the only factor for enhancement of heat transfer.

An extensive study on the heat transfer between supercritical fluid flows and solid walls and also between compressible flows and solid walls is presented in the sixth chapter. It also introduces some practical applications of heat exchangers using supercritical fluid flows and/or compressible flows.

The seventh chapter highlights methods for the design and the choice of heat exchangers to be used in aeronautical applications and provides some practical case studies.

The eighth chapter gives brief overview of the steam generators used in nuclear power plants, their design, operation, and the correlations used to quantify heat transfer in these devices.

The final chapter underlines the importance of corrosion control, fouling cleaning, and enforcement of specific standards for maintenance and cleaning procedures in industries. It also proposes the application of a mitigation approach to deal with fouling and corrosion.

This book is intended to be a useful source of information for researchers, postgraduate students, academics, as well as designers and engineers working in the fields of heat exchangers and related industries.

We would like to thank all the authors for their high-quality contributions and the publishing process manager for providing continuous support, which have made possible the completion of this book.

Finally we would like to express our appreciation to our family members for their continued support and patience during the preparation of this book.

S M Sohel Murshed and Manuel Matos Lopes
University of Lisbon,
Lisbon, Portugal

Chapter 1

Introductory Chapter: Advanced Features and Applications of Heat Exchangers–An Outline

S M Sohel Murshed and Manuel L Matos Lopes

Additional information is available at the end of the chapter

http://dx.doi.org/10.5772/intechopen.68473

1. Introduction

This chapter aims to provide an outline of the various applications of heat exchangers, taking in particular attention related to advanced features as well as to briefly highlight the main aspects from each chapter contribution of this book.

The importance of the advances on the development of new generation heat exchangers and their innovative applications and the subsequent improvement in energy efficient processes relates directly with the energy issues that occupy the first place in the list of the most urgent to solve problems for mankind [1]. It is well accepted that any contribution to improve the way we deal with energy will help in solving our most dramatic problems like water, food, environment, poverty, violent conflicts, disease, education, and population.

Heat exchangers are present in every industrial processes from geothermal and fossil power generation to refrigeration and desalination, in general, as a fundamental part of energy transfer and saving systems. They are also present in our everyday life in all kind of vehicles (terrestrial, nautical and aeronautical), air conditioning, domestic heating and cooling, electronics cooling, and domestic appliances. Heat exchanger technology has been extensively explored and reported in the literature. Nevertheless, there is always the need of more work in this field for a comprehensive understanding of the advances related to edge applications that will contribute decisively to improve the sustainable uses of energy.

In the first of this two book series, "Heat Exchangers—Design, Experiment and Simulation", our attention is centered on the fundamental aspects related to the making of a heat exchanger and its more recent developments [2]. The present volume complements the study of heat exchangers, having under view the aspects related to the innovation in terms

INTECH
open science | open minds

© 2017 The Author(s). Licensee InTech. This chapter is distributed under the terms of the Creative Commons Attribution License (http://creativecommons.org/licenses/by/3.0), which permits unrestricted use, distribution, and reproduction in any medium, provided the original work is properly cited.

of construction and materials and also related to their applications from the most common to the more special cases.

2. Contributions highlight

Each chapter contribution of this book has been briefly highlighted in this section.

The authors Oh and co-workers demonstrated the two-phase flow pressure drop and heat transfer of a refrigerant during boiling in macro-scale and mini-scale channels type heat exchangers. The pressure drop and local heat transfer coefficients were obtained for varying heat and mass fluxes, vapor quality, and saturation temperatures in test sections of various tube diameters. The effects of all these variables on pressure drop and heat transfer coefficient were analyzed. The experimental results were compared against several existing pressure drop and heat transfer coefficient correlations. A new heat transfer coefficient correlation was also proposed that achieved a good agreement with the experimental data.

Having a decade long experience on compact heat exchangers, Erbay and others made a short review on history and basic features on the subject to introduce the offset strip fin geometry. Then, they analyzed the effect of the fin geometry on the performance of the offset strip fin based on experimental and numerical approaches. They demonstrated physical impact on the flow using different offset strip geometries. The thermo-hydraulic features of the flow in the offset strip fin were investigated by considering the Colburn j-factor and friction factors (f) in diverse flow regimes. Furthermore, other criteria derived from the mentioned dimensionless factors were also used as a measure of the performance of the structure.

The same research group leading by Erbay follows in the next contribution with the description of the basic physical features and the analysis of the thermal-hydraulic performance of the heat exchangers with louver fins. They referred the terminology, which is used widely in the field of this kind of compact heat exchanger. They used different flow visualization techniques to study how the flows are affected by the operating conditions of the heat exchanger and the geometric parameters of the louvered fin. A methodology was introduced to calculate the heat transfer coefficient and the friction factor. As in the previous chapter, Colburn j-factor and friction factor and also the Stanton number were defined as performance criteria and once more, the variations of these criteria with respect to Reynolds number and the geometric parameters of the louvered fin were analyzed in terms of the thermo-hydraulic features. The combinations of these dimensionless numbers were discussed in terms of overall performance criteria. Finally, the correlation of the louvered fin heat exchanger and the resulting data were summarized.

The present state of knowledge of free-convection condensation heat transfer on geometrically enhanced tubes was presented by Ali, covering the research on condensate flooding or retention and the experimental as well as theoretical works on geometrically enhanced tubes. Extensive experimental work performed on integral-fin tubes was addressed showing that geometry is not the only point of interest for enhancement of heat transfer. Also a reasonable amount of experimental work was reported on condensation heat transfer on enhanced pin-fin tubes. This showed the superior performance of such tubes over equivalent integral-fin tubes. The extent of condensate retention and formation of many sharp surfaces enhancing surface tension effects on pin-fin tubes was identified to be the important parameters contributing toward the heat transfer enhancement. A previous model by this author was referred and reported to reasonably predict heat transfer on the pin-fin tubes by taking into account the effect of both gravity and surface tension condensate drainage.

Author Ito described the heat transfer between supercritical fluid flows and solid walls and also between compressible flows and solid walls. It started with the explanation of the physical fundamentals of supercritical and compressible fluids. Then, heat transfer performance was obtained by using estimation methods based on the physical fundamentals and conventional experimental results as well as by known correlations. Finally, examples of practical heat exchangers using supercritical fluid flows and/or compressible flows were presented.

Heat exchangers are crucial in the aviation engineering and have a fundamental role especially into reducing the temperatures of the fuel increasing the efficiency of the aircraft engines. The contribution by Carozza explores methods for the design and the choice of heat exchangers to be used with such aeronautical applications and provides some practical case studies. Through this contribution, the author focused on the two main aspects of this class of flow systems, which are widely investigated: fluid flow and heat transfer performances as well as criteria for evaluating those performances. Besides this, several other important aspects related to the need to use a smart and light equipment inside a transport system were discussed. It was stressed that a particular attention should be paid to the selection of components, for example, in the engine zone, not only to reduce the weight but also to improve the whole heat transfer efficiency. Thus, engineers are focusing on new materials, for example, porous materials that have attracted a number of efforts to develop methods suitable to the design and use of such new technologies.

Steam generators used in nuclear power plants are a particular type of heat exchangers. In the steam generators, the heat produced in the reactor core is transferred to the secondary side, the steam supply system, generating the steam to propel the electrical turbine generators. Steam generators have to fulfill special nuclear regulatory requirements regarding their size, selection of materials, pressure loads, and impact on the nuclear power plant safety, among others. The primary side fluid is liquid water at the high pressure, and the fluid on the secondary side is saturated water-steam mixture at the pressure twice as low. A special attention must be given to preserving the boundary between the contaminated water in the primary reactor coolant system and the water-steam mixture in the secondary system. The authors Šadek and Grgić provided a brief overview of the steam generators used in nuclear power plants, its design, operation, and the mathematical correlations used to quantify heat transfer in these devices. Results of the steam generator transient behavior obtained by the simulation with a best-estimate computer code developed for safety analyses of nuclear power plants were also presented. In particular, two types of steam generators were analyzed: the inverted U-tube steam generator which is commonly found in the present-day pressurized water reactors and the helical-coil steam generator that is part of the new generation reactor designs.

In the last contribution, Teng and co-authors proposed a mitigation approach to deal with the major unresolved problems in heat exchanger operation, fouling and corrosion. Here, they consolidated basic background and concepts for the design and operation of heat exchangers to introduce closely related industrial practices for cleaning and green technology maintenance of heat exchangers. For an industry, the proper cleaning method and control play an important role to reduce the production costs. Production cost significantly increases due to chemical usage, maintenance work and downtime loss, and water wastage. Therefore, the authors underlined the importance of corrosion control, fouling cleaning, and enforcement of specific standards for cleaning procedures in the industries. They also proposed the application of a mitigation approach to deal with fouling and corrosion.

3. Other selective complementary sources

A very general review on advances in heat transfer enhancements was performed by Siddique et al. [3]. They addressed most usual heat transfer enhancers: fins and microfins, porous media, large particles suspensions, nanofluids, phase-change devices, flexible seals, flexible complex seals, vortex generators, protrusions, and ultra-high thermal conductivity composite materials. In addition, theoretical enhancement factors along with numerous heat transfer correlations were presented in this review for each heat transfer enhancer.

3.1. Materials

The conventional heat exchangers are manufactured in metal (such as stainless steel, copper and aluminum) and have disadvantages in terms of weight and cost. In addition, specially treated metal heat exchangers are needed if the working fluids are corrosive. Alternative materials have been used for heat exchangers that can overcome some of the disadvantages of the conventional ones like weight, cost, and chemical resistance, be more adequate for particular applications, and also have comparable heat exchange efficiency and be easily fabricated.

Due to their low cost, lightweight, and corrosive resistant features, *polymer* heat exchangers are receiving growing interest from researchers, engineers, and other industry related players. Thus, as a better alternative to metallic heat exchangers in a wide range of applications, these particular heat exchangers have extensively been investigated in recent years. This can be evidenced from a recent review paper by Chen et al. [4] who reported developments including theoretical modeling, experimental findings, heat transfer enhancement methods of polymer materials, and a wide range of applications of polymer heat exchangers. Another interesting review work by T'Joen et al. [5] discussed the use of polymer matrix composites in HVAC&R applications, showing how a careful material selection and modification of the design allows to fully exploit the material properties.

Lin et al. [6] proposed a new compact *graphite foam* heat exchanger for vehicle cooling application that allows to match the increasing cooling power and space limitation in vehicles. Their simulation results show that the wavy corrugated foam presents high thermal performance and low pressure drop. A comparative study between the wavy corrugated foam heat exchanger and a conventional aluminum louver fin heat exchanger was also carried out to evaluate the performance of graphite foam heat exchangers. Furthermore, several recommendations were made about the further development of the application of graphite foam heat exchangers in vehicles.

For heat exchanger applications needing extreme operation temperatures such as in the field of power generation or heat recovery, *ceramics* and ceramic matrix composites are suitable to design heat exchangers and particularly adequate to attain optimal overall efficiency, cost, and size of the system. The review by Sommers et al. [7] provides the current state-of-the-art of ceramic materials for use in a variety of heat transfer systems.

3.2. Nanofluids

Recent advances in nanotechnology have allowed the development of a new category of fluids termed "nanofluids" [8]. Among many other applications, nanofluids can be used as thermal fluids and are considered heat transfer enhancers. The review by Huminic and Huminic [9] summarized the important available publications on the enhancement of the convection heat transfer in heat exchangers using nanofluids. They also presented the theoretical and experimental results for the effective thermal conductivity, viscosity, and the Nusselt number reported in the literature. It also focused on the application of nanofluids in various types of heat exchangers: plate, shell and tube, compact, and double-pipe heat exchangers.

A special type of nanofluids is ionanofluids that can be defined as suspensions of nanomaterials in ionic liquids. Ionic liquids possess promising thermophysical properties and great potential for numerous applications, particularly as new heat transfer fluids. Since ionic liquids are the base fluids in ionanofluids, the above-mentioned heat transfer enhancement obtained with nanomaterials suspensions can be potentiated by the thermophysical properties of ionic liquids. Nieto de Castro et al. [10] presented some pioneering researches that indicate that ionanofluids show great promises to be used as innovative heat transfer fluids in heat exchangers and novel media for many green energy-based applications.

3.3. Special design

The efficient design of heat exchangers is more critical in some applications, requiring devices having superior performance and reliable mechanical characteristics at high pressure and high temperature and complying with geometric constraints. Therefore, design of heat exchangers is one of the aspects where continuous advances are registered.

The review by Li et al. [11] reported the performances of *compact* heat exchangers, including well-established devices, some relative newcomers to the market and also designs still being tested in the laboratory. The structures, heat transfer enhancement mechanisms, advantages, and limitations are summarized, and an example of an application as a solar receiver is given. It also referred available correlations for heat transfer and friction factor developed by various researchers.

Microchannels represent the next step in heat exchanger development due to their high heat transfer performance and reduced weight as well as their space, energy and materials savings potential. Khan and Fartaj [12] made a survey of the published literature on the status and potential of microchannels, identifying research needs. They also developed an experimental infrastructure to investigate the heat transfer and fluid flow for a variety of working fluids in different microchannel test specimens.

The study by Abu-Khader [13] presented the advances in *plate* heat exchangers both in theory and applications. The selected areas discussed in this review are the ones that attracted more attention recently, namely compactness and downsizing without the loss of performance, which is crucial for the industry; theoretical developments; reducing fouling and corrosion of plates in sever processes, with a direct impact on operational cost; and using nanofluids.

In the paper by Wang et al. [14], a general review was provided on developments and improvements of *shell-and-tube* heat exchangers with helical baffles of different improved designs. Extensive results from experiments and numerical simulations indicated that these heat exchangers have better flow and heat transfer performance than the conventional baffled heat exchangers, therefore allowing to save energy, reduce cost, and prolong service life and operation time in industrial applications.

The heat pipes are accepted as an excellent way of saving energy due to the high heat recovery effectiveness of these devices. A brief literature review was performed by Srimuang and Amatachaya [15] on the applications of *heat pipes* heat exchangers for waste heat recovery in both commercial and industrial applications. The authors also summarized the experimental studies on the conventional heat pipe, two-phase closed thermosiphon, and oscillating heat pipe heat exchangers.

3.4. Applications

There are review studies available on other types of heat exchangers used in particular cases of applications where innovative and state-of-the-art equipment must be developed and used, such as exemplified with the cases of geothermal processes with ground heat exchangers [16] and biotechnological industries [17].

4. Conclusions

This chapter briefly discusses the importance of heat exchangers and their advanced applications in terms of energy efficiency and process intensification, minimizing environmental and societal impacts. It highlights main research and findings from each contributed chapter of this book. It also provides key topics related to advanced features and applications of heat exchangers and corresponding reference sources. We believe that this book will be a useful reference source of information on advanced features and applications of heat exchangers.

Author details

S M Sohel Murshed* and Manuel L Matos Lopes

*Address all correspondence to: smmurshed@ciencias.ulisboa.pt

University of Lisbon, Lisbon, Portugal

References

[1] WEO2016 – World Energy Outlook 2016 (released on 16 November 2016). International Energy Agency, Available from: http://www.worldenergyoutlook.org/publications/weo-2016/ (consulted on March 8, 2017)

[2] Murshed SMS, Matos Lopes ML. Heat Exchangers-Design, Experiment and Simulation. Rijeka: INTECH; 2017

[3] Siddique M, Khaled ARA, Abdulhafiz NI, Boukhary AY. Recent advances in heat transfer enhancements: A review report. International Journal of Chemical Engineering. 2010;**2010**:28. Article ID 106461

[4] Chen X, Su Y, Reay D, Riffat S. Recent research developments in polymer heat exchangers – A review. Renewable and Sustainable Energy Reviews. 2016;**60**:1367–1386

[5] T'Joen C, Park Y, Wang Q, Sommers A, Han X, Jacobi A. A review on polymer heat exchangers for HVAC&R applications. International Journal of Refrigeration. 2009;32: 763–779

[6] Lin W, Sundén B, Yuan J. A performance analysis of porous graphite foam heat exchangers in vehicles. Applied Thermal Engineering. 2013;**50**:1201–1210

[7] Sommers A, Wang Q, Han X, T'Joen C, Park Y, Jacobi A. Ceramics and ceramic matrix composites for heat exchangers in advanced thermal systems-A review. Applied Thermal Engineering. 2010;30:1277–1291

[8] Murshed SMS, Nieto de Castro CA. Nanofluids: Synthesis, Properties and Applications. New York: Nova Science Publishers; 2014

[9] Huminic G, Huminic A. Application of nano fluids in heat exchangers: A review. Renewable and Sustainable Energy Reviews. 2012;16:5625–5638

[10] Nieto de Castro CA, Murshed SMS, Lourenço MJV, Santos FJV, Matos Lopes ML, França MP. Ionanofluids – New heat transfer fluids for green processes development, Chapter 18 in Green Solvents I. A. Mohammad, Inamuddin, editors. Properties and Applications in Chemistry.Dordrecht: Springer; 2012.

[11] Li Q, Flamant G, Yuana X, Neveu P, Luod L. Compact heat exchangers: A review and future applications for a new generation of high temperature solar receivers. Renewable and Sustainable Energy Reviews. 2011;**15**:4855–4875

[12] Khan MG, Fartaj A. A review on microchannel heat exchangers and potential applications. International Journal of Energy Research. 2011;**35**:553–582

[13] Abu-Khader MM. Plate heat exchangers: Recent advances. Renewable and Sustainable Energy Reviews. 2012;**16**:1883–1891

[14] Wang Q, Chen G, Chen Q, Zeng M. Review of improvements on shell-and-tube heat exchangers with helical baffles. Heat Transfer Engineering. 2010;**31**(10):836–853

[15] Srimuang W, Amatachaya P. A review of the applications of heat pipe heat exchangers for heat recovery. Renewable and Sustainable Energy Reviews. 2012;**16**:4303–4315

[16] Florides G, Kalogirou S. Ground heat exchangers—A review of systems, models and applications. Renewable Energy. 2007;**32**:2461–2478

[17] Agarwal P, Sikand A, Shanthi V. Application of heat exchangers in bioprocess industry: A review. International Journal of Pharmacy and Pharmaceutical Sciences. 2014;**6**:24–28

Chapter 2

Pressure Drop and Boiling Heat Transfer Characteristics of R410A in Macro-Scale and Mini-Scale Channels

Jong-Taek Oh, Nguyen Ba Chien, Kwang-Il Choi and
Pham Quang Vu

Additional information is available at the end of the chapter

http://dx.doi.org/10.5772/65966

Abstract

This chapter demonstrates the two-phase flow pressure drop and heat transfer of R410A during boiling in various tube types. The pressure drop and local heat transfer coefficients were obtained for heat fluxes ranging from 10 to 40 kW/m^2, mass fluxes ranging from 100 to 600 kg/m^2s, the vapour quality up to 1.0 and the saturation temperatures of 5–15°C. The test sections were made of various tube diameters of 1.5, 3.0, 6.61 and 7.49 mm, respectively. The effect of mass flux, heat flux, saturation temperature and inner tube diameter on pressure drop and heat transfer coefficient was analysed. The experimental results were compared against several existing pressure drop and heat transfer coefficient correlation. New correlations of pressure drop and boiling heat transfer coefficient were also developed in this present study.

Keywords: mini-channels, heat transfer, pressure drop, correlation, heat exchanger

1. Introduction

Mini-channels are progressively used in making the compact heat exchangers nowadays. The application of these heat exchanger types in refrigeration and air-conditioning fields shows various advantages such as higher efficiency, lower air side pressure drop, reducing refrigerant charge and the more compactness size compared to the conventional types. The development of numerous tube types enables the creation of even more effective compact heat exchangers.

INTECH
open science | open minds

© 2017 The Author(s). Licensee InTech. This chapter is distributed under the terms of the Creative Commons Attribution License (http://creativecommons.org/licenses/by/3.0), which permits unrestricted use, distribution, and reproduction in any medium, provided the original work is properly cited.

At the low value of saturation temperature (normally from -30 to 20°C), say, those are typically applied in refrigeration and air-conditioning, Kim et al. [1] investigated the boiling heat transfer coefficient of R-410A in smooth/micro-fin tubes within the conditions: the saturation temperature ranged from -15 to 5°C, the mass fluxes of 70–211 kgm^2/s and the heat fluxes of 5–15 kW/m^2. In this study, the authors reported that the heat transfer coefficients increased with the increasing of heat flux and mass flux. In addition, the average heat transfer coefficients of micro-fin tubes were 80–150% and 10–60% higher than those of smooth tubes for the outside diameter (OD) of 9.52 and 7.0 mm, respectively. Kim et al. [2] demonstrated the boiling heat transfer of R-410A in horizontal copper tubes. The results were carried out in 9.52 mm OD tube with following conditions: the saturation temperature of 15°C, the heat flux of 11kW/m^2 and the mass flow of 30–60 kg/h. This study also reported that the average evaporation heat transfer coefficients of micro-fin tubes were higher than those of the smooth tubes for both refrigerants, R22 and R-410A. In the other research, Wellsandt et al. [3] reported the heat transfer coefficient and pressure drop of R410A and R407C during evaporation inside horizontal herringbone-micro-fin tubes. The authors found that, at moderate vapour quality regime, the effect of mass flux on the heat transfer coefficient was insignificant while a strong influence was observed when the vapour quality was over 60%. Inside smooth tube, the heat transfer coefficient, pressure drop and flow pattern of CO_2, R410A and R22 were investigated by Park et al. [4]. The results were reported for 6.1 mm inner diameter tube within the saturation temperature of -15 and -30°C, the mass flux of 100–400 kgm^2/s, the heat flux of 5–15 kW/m^2 and the vapour quality ranged from 0.1 to 0.8. In this study, the heat transfer coefficients of R-410A are affected by the change of heat flux, mass flux and quality. The nucleate and convective boiling heat transfer mechanisms were consequently activated.

For high evaporation temperature application such as industrial heat pump systems, Padovan et al. [5] presented the experimental results of boiling heat transfer of R134a and R410A in horizontal micro-fin tube at high saturation temperature. The wide range of testing conditions that were investigated include: the mass flux from 80 to 600 kgm^2/s, the heat flux from 14 to 83.5 kW/m^2, the vapour quality from 0.1 to 0.99 and covered the saturation temperature of 30 and 40°C. The dominance of convective boiling mechanism on heat transfer coefficient was observed at the saturation temperature of 30°C. Moreover, the effect of nucleate boiling mechanism was more distinct when the saturation temperature raised. On the other side, the heat transfer coefficient of R410A in micro-fin tube is higher than that in plain tube. This result agreed well with the others in previous studies.

In addition, beside the researches on single channel, various studies on heat transfer and pressure drop of R410A in multi-port tube have been published. Cavallini et al. [6] reported the frictional pressure gradient of R236ea, R134a and R410A inside multi-port mini-channels with the hydraulic diameter of 1.4 mm. The experimental results covered a wide range of reduced pressure from 0.1 to 0.5, and the mass flux ranged from 200 to 1400 kgm^2/s. The study showed that the existing frictional pressure drop correlations were unable to predict the data of R410A. Jatuporn et al. [7] investigated the heat transfer coefficient and pressure drop of R410A in horizontal aluminium multi-port mini-channel having the hydraulic diameter of 3.48 mm. The strong effects of mass flux, heat flux and saturation temperature on heat transfer coefficient were observed while only mass flux and saturation temperature affected on pressure drop. Recently, Chien et al. [8] also reported the heat transfer and pressure drop of R410A in multi-port mini-channels with smaller hydraulic diameters, 1.14 and 1.16 mm. The data were conducted with the mass fluxes of 50–150 kgm^2/s, the heat fluxes of 3 and 6 kW/m^2 and the saturation temperature of 6°C. This study reported that only heat flux affected on heat transfer coefficient of R410A while only mass flux affected on pressure drop. A heat transfer coefficient correlation was also developed in this study.

However, due to the variation in tube geometry and diameter, the heat transfer characteristics and pressure drop of R410A during evaporation process should be experimentally validated to optimise the design of heat exchangers. In order to bring out an overview to the reader, this chapter investigates the experimental results with the wide range of operating conditions as well as the tube diameters, which have been evaluated in our lab in the past [9–11] and recently. The influence of mass flux, heat flux and channel diameter on the heat transfer coefficient and pressure drop was well reported. In addition, the comparison between the experimental results and several existing pressure drop and heat transfer coefficient correlations was carried out. Finally, the development of correlations of pressure drop and heat transfer coefficient correlations for heat exchanger with mini-channel design was demonstrated in this study.

2. Experimental facilities and data reduction

2.1. Experimental model

Figure 1(a) and **(b)** depicts the schematic diagram of the experimental apparatus for copper tube and stainless tube, respectively. In general, both facilities mainly consist of a refrigerant loop, the water loops and the data acquisition system. The refrigerant loop included the receiver tank, the sub-cool unit, the Coriolis mass flow meter, the condense unit and test sections. In boiling mode, the refrigerant was delivered clockwise by the gear pump. The mass flow rate could be controlled by changing the pump speed and was measured by the flow meter. The quality of flow was adjusted in pre-heater to the desired value before entering the test section. The test sections could be heated by a water loop or an electric transformer as shown in the figure. When using the water loop, the heat capacity could be varied by mastering the mass flow rate or temperature of inlet water. On the other hand, when using the transformer, the power could be set by controlling the input voltage. The refrigerant was then evaporated in the test section. The vapour at the outlet of test section was condensed by a condenser unit and accumulated into the receiver. The experimental apparatus was well insulated with rubber and foam to minimise the effect of surround environment temperature.

The detail of test sections for copper tube was depicted in **Figure 2(a)**. The test tubes were made from copper with the inner diameters of 6.61 and 7.49 mm. The effective length was 1200 mm. As shown in the figure, the test sections were divided into four separated subsections with a length of 300 mm. Along the test section, the T-type thermocouples were attached, each 150 mm of length, at three positions: top, middle and bottom. At the adiabatic pipes between two consecutive sub-sections, thermocouples and pressure transducers were

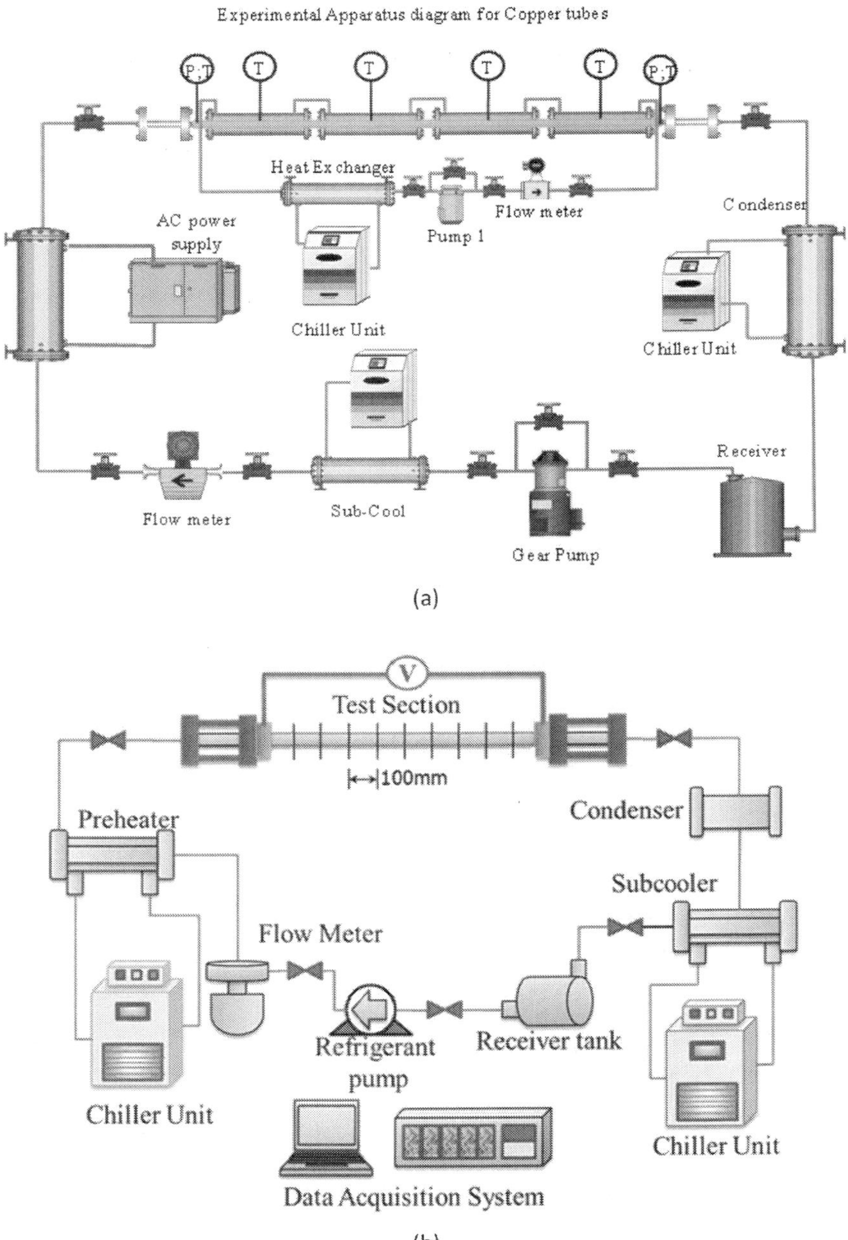

Figure 1. The experimental apparatus: (a) copper tube and (b) stainless tube.

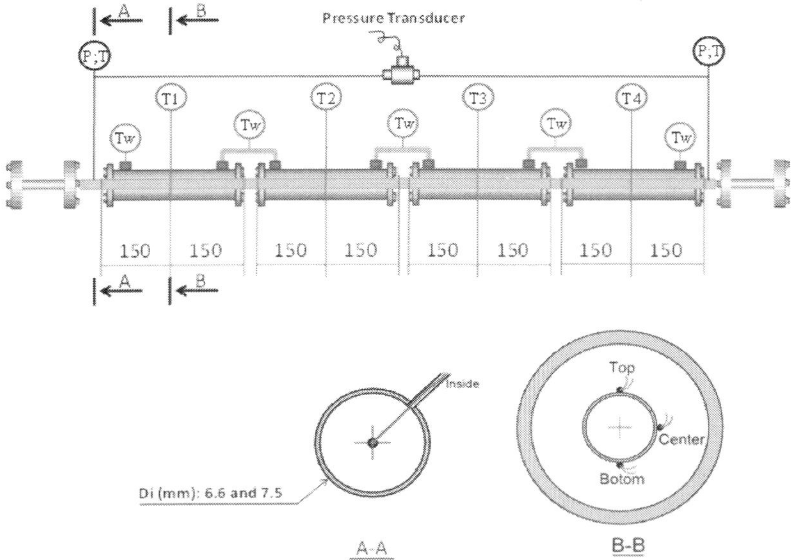

(a)

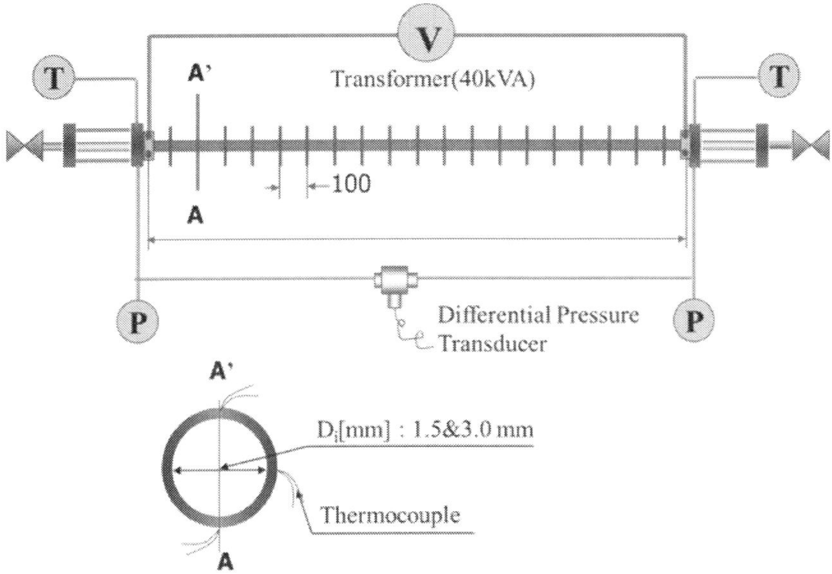

(b)

Figure 2. The test sections: (a) copper tube and (b) stainless tube.

set up to measure the local pressure drop. Two sight glasses were installed at the beginning and end of test sections to visualise the flow pattern.

Figure 2(b) describes the detail setup of test section using the stainless tube. The tubes have a diameter of 1.5 and 3.0 mm with a length of 1500 and 3000 mm, respectively. The T-type thermocouples were also attached at the top, middle and bottom points, each 100 mm, along the test section from the inlet. Two pressure transmitters were located at the inlet and outlet of the test section. To enhance the accuracy of pressure drop value, a differential pressure transducer was also set up.

The physical properties of the refrigerant were obtained from REFPROP 8 [12]. The temperature, pressure and mass flow rate data were recorded using the data acquisition.

2.2. Data reduction

The data were collected using a data acquisition and were analysed in real time by the data reduction program. All the information about test conditions and data during the operation were displayed on the monitor.

2.2.1. Heat transfer coefficient

In the case of using water heating loop, the heat capacity of each sub-section Q_n is calculated from mass flow rate and rising enthalpy of cooling water flowing inside the water tubes as follows:

$$Q_n = W_n c_p (T_{\text{nout}} - T_{\text{nin}}) - Q_{\text{loss}} \tag{1}$$

where $\dot{m}_n$, c_p, and T_n are the mass rate of flow, specific heat and temperature of cooling water in sub-section n, respectively. Q_{loss} is the heat loss on the test section, which was determined when calibrating the system. In the case of using an electric transformer, the heat capacity can be determined by the root mean square values of electric voltage and current as follows:

$$Q = I_{\text{RMS}} \times U_{\text{RMS}} \tag{2}$$

The local heat transfer coefficient inside the channel can be evaluated as the ratio of the heat flux to saturation minus the inside wall temperature:

$$h = \frac{Q}{A \times (T_{\text{wi}} - T_{\text{sat}})} \tag{3}$$

At the inlet of test section, the saturation temperatures were determined by the local measured pressures. At the outlet of each sub-section, the saturation temperatures were then calculated by subtracting the inlet temperature to the one raised by pressure drop. In the current study, the difference of temperature between the saturated refrigerant and inside wall was very high. In addition, the temperature drop raised by the pressure drop was kind of small. Therefore, the saturation temperature between the inlet and outlet of test section could be calculated as the linear function of two known values. The temperature of inside tube wall, T_{wi}, was determined based on the steady-state one-dimensional radial conduction heat transfer through the wall with and without the internal heat generation for the cases, heat flux was applied by the electric transformer and hot water, respectively.

The vapour quality, x, at the measurement locations, z, was determined based on the thermodynamic properties:

$$x = \frac{i - i_f}{i_{fg}} \tag{4}$$

In some cases, since the refrigerant flow could not be saturated completely before entering the test section, the sub-cooled length was determined as follows:

$$z_{sc} = L\frac{i_f - i_{fi}}{\Delta i} = L\frac{i_f - i_{fi}}{(Q / W)} \tag{5}$$

2.2.2. Pressure drop

The total pressure drop of two-phase fluid in general is calculated by the sum of the static head Δp_{static}, the momentum $\Delta p_{\text{momentum}}$ and the frictional pressure drop $\Delta p_{\text{frictional}}$:

$$\Delta p_{\text{total}} = \Delta p_{\text{static}} + \Delta p_{\text{mom}} + \Delta p_{\text{frict}} \tag{6}$$

Since the horizontal tube was used in this study, the pressure head can be neglected. The saturated refrigerant from liquid is evaporated linearly with the test distance to vapour-liquid mixture at the vapour quality x. Therefore, the momentum pressure drop can be defined by following equation:

$$-\left(\frac{dp}{dz}a\right) = G^2 \upsilon_f \left\{ \left[\frac{x^2}{\alpha}\left(\frac{\upsilon_g}{\upsilon_f}\right) + \frac{(1-x)^2}{1-\alpha} \right]_{\text{out}} - \left[\frac{x^2}{\alpha}\left(\frac{\upsilon_g}{\upsilon_f}\right) + \frac{(1-x)^2}{1-\alpha} \right]_{in} \right\} \tag{7}$$

For horizontal tube, void fraction α is defined by Steiner [13] equation that was modified from Rouhani-Axelsson [14] model:

$$\alpha = \frac{x}{\rho_g}\left[(1 + 0.12(1-x))\left(\frac{x}{\rho_g} + \frac{1-x}{\rho_f} \right) + \frac{1.18(1-x)\left[g\sigma(\rho_f - \rho_g) \right]^{0.25}}{G\rho_f^{0.5}} \right]^{-1} \tag{8}$$

3. Result and discussions

3.1. Pressure drop

Figure 3 illustrates the significant effect of mass flux and vapour quality on pressure drop gradient of copper tube with the inner diameters of 6.61 and 7.49 mm. The pressure drops are higher with the higher mass flux. The reason can be explained in that, in horizontal tube, the

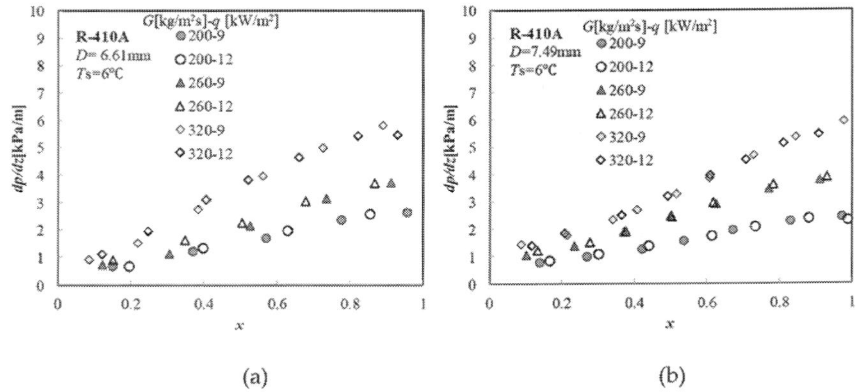

Figure 3. The effect of mass flux, heat flux and mass quality on pressure drop of R410A: (a) Di = 6.61 mm; (b) Di = 7.49 mm.

pressure drop is mainly contributed by the frictional and momentum pressure components. When the mass flux increases, the flow velocity increases, which raises both of two components. This effect of mass flux on the two-phase pressure drop was well observed in various studies including the single-mini/conventional-channel and multi-port tubes [4, 8, 15–19].

In addition, the pressure drop gradient also increases with the increase of mass quality. When the dry out occurred, the reduction in frictional pressure drop was observed. Moreover, the effect of heat flux on pressure drop is unclear.

The effects of tube diameter are depicted in **Figure 4**. The pressure drops of gradient increases when the tube diameter decreases. The wall shear stress that raises higher in smaller tube is believed to be the reason for this phenomenon.

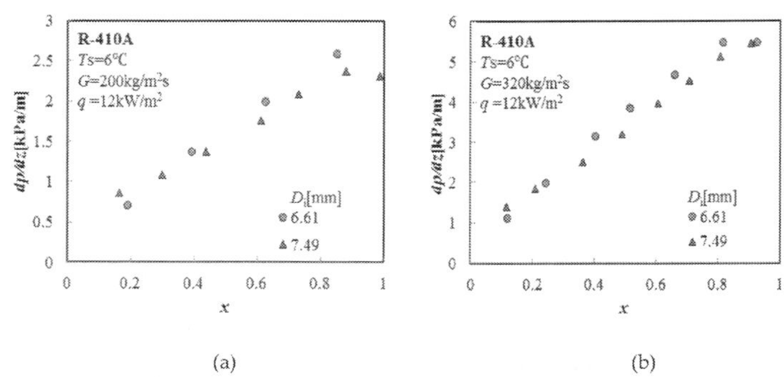

Figure 4. The effect of inner tube diameter on pressure drop of R410A: (a) G = 200 kg/m2s; (b) G = 320 kg/m2s.

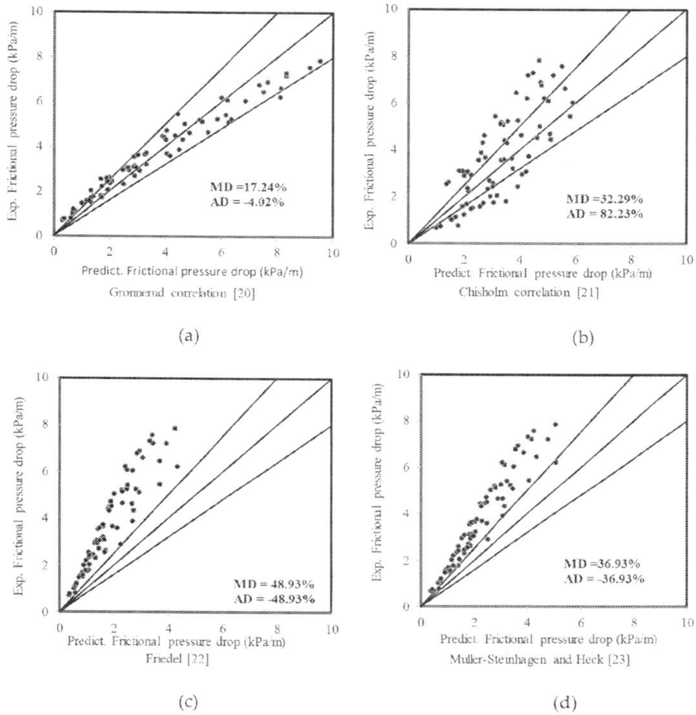

Figure 5. Comparison of experimental data with some existing frictional pressure drop correlations: (a) Gronnerud [20]; (b) Chisholm [21]; (c) Friedel [22]; (d) Muller-Steinhagen and Heck [23].

In this study, the experimental data are validated by comparing with some well-known frictional pressure drop correlations [20–23], as shown in **Figure 5**. Among them, a correlation proposed by Gronnerud [20] shows the best prediction with a mean deviation of 17.24%. His correlation determined the dependence of frictional pressure drop on the Froude number and could be applied to a wide range of mass quality from 0 to 1. Both of two correlations proposed by Chisholm [21] and Muller and Heck [23] have mean deviations of above 30%. The correlation proposed by Chisholm [21] was based on the empirical method and the one proposed by Muller and Heck [23] expressed the interpolation between the liquid and vapour flow. The correlation proposed by Friedel [22] used a wide range of data and was commended for the ratio of liquid/vapour, which is less than 1000.

In order to improve the prediction of pressure drop of R410A in horizontal tube as well as simplify the calculation, a new frictional pressure drop is developed. The general form is developed based on the ideal of Lockhart-Martinelli [24]. The frictional pressure drop is comprised of frictional pressure drop of liquid-phase adjusted by two-phase multiplier. The equation can be determined as follows:

$$\frac{dp}{dz}f = \left(\frac{dp}{dz}F\right)_{fo}\phi_f^2 \tag{9}$$

where the frictional pressure drop of liquid phase is defined as follows:

$$\left(\frac{dp}{dz}F\right)_{fo} = \frac{2f_{fo}G^2}{D_h\rho_f} \tag{10}$$

and f_{f_0} is frictional factor, which can be determined based on the flow regime:

$$\begin{aligned}
&\mathrm{Re}_{fo} < 2300 : f_{\text{laminar}} = \frac{16}{\mathrm{Re}_{fo}} \\
&\mathrm{Re}_{fo} > 3000 : f_{\text{Blasius}} = 0.079\,\mathrm{Re}_{fo}^{-0.25} \\
&2300 < \mathrm{Re}_{fo} < 3000 : \\
&f = \frac{f_{\text{blasius}} - f_{\text{laminar}}}{3000 - 2300}(\mathrm{Re}_{fo} - 2300) + f_{\text{laminar}}
\end{aligned} \tag{11}$$

From that, a regression program is developed using the experimental data as the input variable. The final factor is calculated and expressed as follows:

$$\phi_f^2 = 62.373x^{1.086}(1-x)^{0.151} \tag{12}$$

The proposed frictional pressure drop correlation shows a good prediction with the mean deviation of 9.29% and the average deviation of -0.69%. The comparison between the experimental data and predicting model is shown in **Figure 6**.

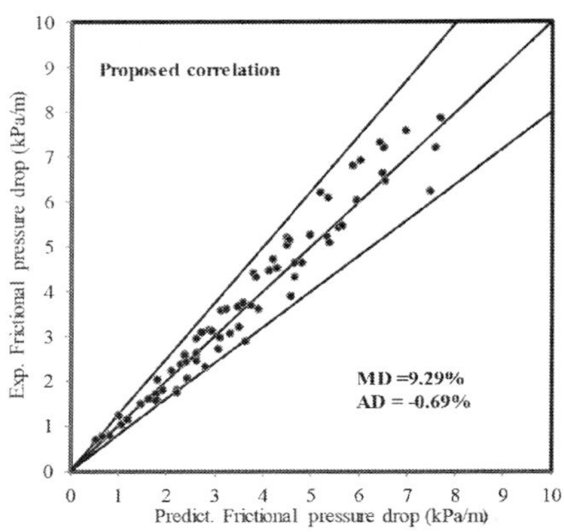

Figure 6. Comparison of experimental data with proposed frictional pressure drop correlation.

3.2. Heat transfer coefficient

Figure 7(a) and **(b)** illustrates the effect of mass flux on the heat transfer coefficient of R410A in 1.5 and 6.61 mm inner diameter, respectively. In low quality regime, the effect of mass flux in 1.5 mm tube, called mini-channel, is insignificant while the inverse trend of mass flux effect is shown in 6.61 mm tube, called conventional channel. It indicates that the nucleate boiling is predominant in mini-channel. The similar results are reported in other studies [10, 25–28]. The different contribution of nucleate boiling to the heat transfer coefficient of refrigerant between mini-channel and conventional channel, therefore, should be validated before applying the heat transfer model of macro-scale to mini-scale channel and vice versa. At higher quality regime, due to the active force of convective boiling, the effect of mass flux on heat transfer coefficient is clear. The higher mass flux raises the higher heat transfer coefficient.

The strong effect of heat flux on the heat transfer coefficient is shown in **Figure 8**. The trend depicts that the heat transfer coefficient of R-410A was higher with the increase in heat flux and vapour quality. The similar phenomenon was also observed in the studies of Park et al. [4] and Mastrullo et al. [29]. That means the convective boiling mechanism was active on the heat transfer mechanism of R-410A in mini-channel.

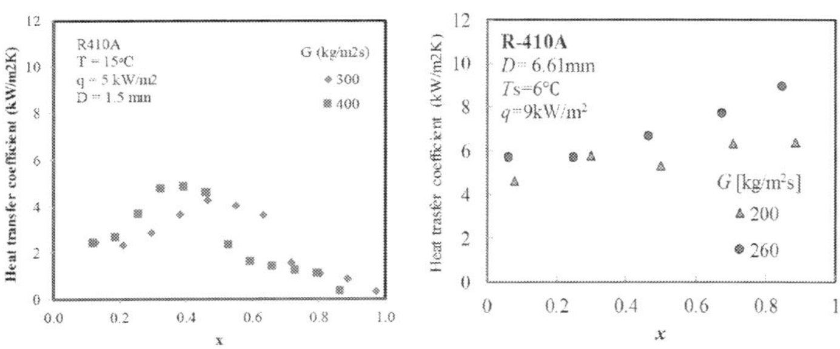

(a) (b)

Figure 7. The effect of mass flux on heat transfer coefficient of R410A: (a) Di = 1.5 mm; (b) Di = 6.61 mm.

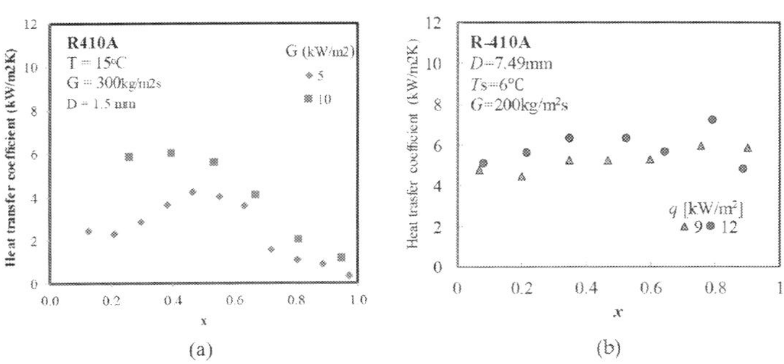

(a) (b)

Figure 8. The effect of heat flux on heat transfer coefficient of R410A : (a) Di = 1.5 mm; (b) Di = 7.49 mm.

Figure 9 shows the effect of tube diameter on the heat transfer coefficient. The heat transfer coefficient of R410A was higher with smaller tube diameter. It can be explained that the heat transfer surface area per unit volume increases when the diameter decreases at the same operating conditions, therefore, increase the heat transfer coefficient.

In order to validate the experimental heat transfer coefficient of R410A in both mini- and macro-scale channel, Chien et al. [11] compared the data with various heat transfer coefficient correlations [26, 30–34] include both of the correlations developed for mini-scale channel and the ones developed for macro-scale channel. The data bank consists of 452 data points including 61 data with $D_h > 3.0$ mm (macro-scale) and 391 data with $D_h \leq 3.0$ mm (mini-scale). The distribution of data is depicted in **Figure 10** and the comparison is depicted in **Figure 11**. The summary of heat transfer coefficient correlation is depicted in **Table 1**. The inaccuracy of existing general heat transfer correlations in predicting the experimental data of R410A resulted due to the extrapolation of the operating condition from the previous model to the

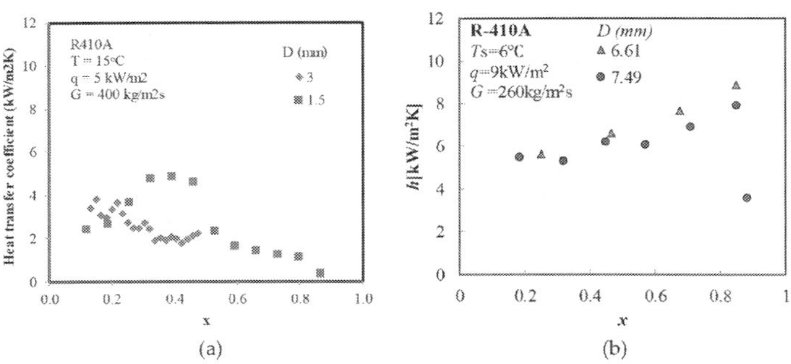

Figure 9. The effect of inner diameter on heat transfer coefficient of R410A : (a) Di = 1.5 & 3.0 mm; (b) Di = 6.61 & 7.49 mm.

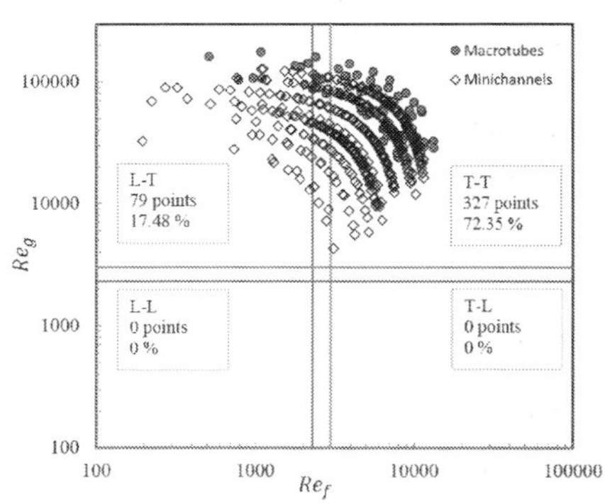

Figure 10. The data distribution.

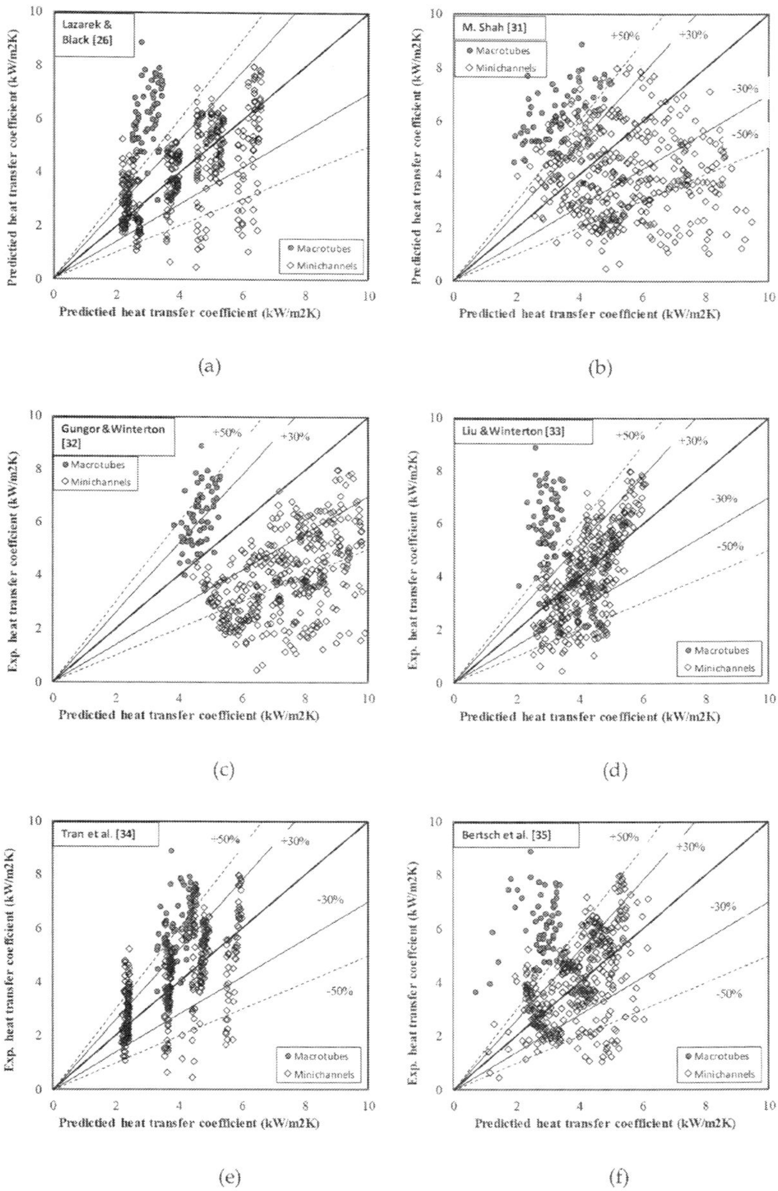

Figure 11. Comparison of experimental data with some existing correlation: (a) Lazarek and Black [26]; (b) Shah [32]; (c) Gungor and Winterton [32]; (d) Liu and Winterton [33]; (e) Tran et al. [34]; (f) Bertsch et al. [35].

1. Lazarek and Black [26] correlation	$h_{tp} = (30\mathrm{Re}_{lo}^{0.857} Bo^{0.714})(\frac{k_l}{D_h});$ $Re_{lo} = \frac{GD_h}{\mu_l}; Bo = \frac{\phi}{Gi_{lg}}; \phi = 1.4\times10^4 - 3.8\times10^5 W/m^2$
2. Shah [30] correlation	$\psi = \frac{h_{TP}}{h_l};\ Co = \left(\frac{1}{x-1}\right)^{0.8}\left(\frac{\rho_v}{\rho_l}\right)^{0.5}$ $Bo = \frac{q}{Gh_{lv}};\ Fr_L = \frac{G^2}{\rho_l^2 gD}$ $h_l = 0.023\mathrm{Re}_l^{0.8}\mathrm{Pr}_l^{0.4}\left(\frac{k_l}{D_i}\right)$ $h_{tp} = \mathrm{Max}(F,S)h_l;$
3. Gungor and Winterton [31] correlation	$h_{tp} = F.h_l + S.h_{nb}$ $h_l = 0.023\mathrm{Re}_l^{0.8}\mathrm{Pr}_l^{0.4}\left(\frac{k_l}{D_i}\right); F = 1 + 24000Bo^{1.16} + 1.37\left(\frac{1}{X_{tt}}\right)^{0.86}$ $h_{nb} = 55P_r^{0.12}q^{2/3}(-\log_{10}p_r)^{-0.55}M^{-0.5}; S = (1+0.55F^{0.1}\mathrm{Re}_{lo}^{0.16})^{-1}$ If the tube is horizontal and the Froude number Fr_l is less than 0.05 then F, S should be multiplied by the factor $F.Fr_l^{(0.1-2Fr_l)}; SFr_l^{0.5}$, respectively.
4. Liu and Winterton [32]	$h_{tp} = \left[(Fh_l)^2 + (Sh_{pool})^2\right]^{0.5},$ $\mathrm{h}_l = 0.023\mathrm{Re}_{lo}^{0.8}\mathrm{Pr}_l^{0.4}\frac{k_f}{D_h}; F = \left[1 + x\mathrm{Pr}_l\left(\frac{\rho_l}{\rho_v} - 1\right)\right]^{0.35}$ $h_{pool} = 55P_r^{0.12}q^{2/3}(-\log_{10}p_r)^{-0.55}M^{-0.5}; S = \left(1+0.55F^{0.1}\mathrm{Re}_l^{0.16}\right)^{-1}$
5. Tran et al. [33]	$h = (8.4x10^{-5})(Bo^2 We_l)^{0.3}\left(\frac{\rho_l}{\rho_v}\right)^{-0.4}$
6. Bertsch et al. [34]	$h_{tp} = S.h_{nb} + F.h_{conv,tp}; h_{nb} = 55P_r^{0.12}q^{2/3}(-\log_{10}p_r)^{-0.55}M^{-0.5};$ $h_{conv,tp} = h_{conv,l}.(1-x) + h_{conv,v}.x;$ $S = 1 - x; F = 1 + 80.(x^2 - x^6).e^{-0.6C_f}$

Table 1. Flow boiling heat transfer correlations considered in this work.

present data as well as the significant effect of specific working fluid as proposed by Kandlikar [35, 36]. The development of a new boiling heat transfer coefficient correlation for R410A in mini- and macro-scale channels, therefore, is needed.

In general, the boiling heat transfer coefficient mainly consists of two mechanisms: the nucleate boiling and the forced convective boiling mechanism. Hence, the form proposed by Chen [37] is widely accepted and is used in this study to develop the new correlation. This form is based on the superposition model with the contribution of two heat transfer mechanisms. The formula is defined as:

$$h_{tp} = F.h_{lo} + S.h_{\mathrm{pool}} \tag{13}$$

In the above equation, the enhancement of convective boiling due to the increase of flow velocity when the vapour quality increases was counted by the enhanced factor F. On the other hand, the suppression of nucleate boiling term when the fluid layer thickness decrease as the vapour quality increases is defined as suppression factor S. The heat transfer coefficient of liquid phase in force convective heat transfer mechanism can be defined by the Dittus-Boelter [38] correlation as follows:

$$h_{lo} = 0.023 \mathrm{Re}_l^{0.8} \mathrm{Pr}_l^{0.4} \frac{k_l}{D} \tag{14}$$

Due to the higher velocities and the development of the thinner film layers, the heat transfer coefficients of two-phase flow are normally higher than those of single phase. For that reason, the enhancement factor F is consequently determined under the effect of the density ratio and vapour quality ratio. In the studies of Chen [37] and Gungor-Winterton [31], enhancement factor F is defined as the function of Martinelli parameter X_{tt}. Shah [30], on the other hand, replaced this parameter by convection number C_o since the influence of viscosity ratio was not found. In present study, the correlate analysis shows the same value of the Martinelli parameter and convection number on the enhancement factor. Therefore, the calculation of F can be simplified by the function of convection number $F = f(C_o)$, where the convection number was defined as follows:

$$C_o = \left(\frac{1-x}{x}\right)^{0.8} \left(\frac{\rho_g}{\rho_l}\right)^{0.5} \tag{15}$$

Using the regression method, the final form of enhancement factor F is proposed as follows:

$$F = 1.061 \exp\left(\frac{0.042}{C_o}\right) \tag{16}$$

Copper [39] developed the heat transfer correlation for nucleate boiling with the large data bank and showed a good prediction. Hence, in the remaining term, the nucleate boiling and the Copper correlation can be used to determine the pool boiling heat transfer. The equation is defined as follows:

$$h_{\mathrm{pool}} = 55 P_R^{0.12} (-\log_{10}(P_R))^{-0.55} M^{-0.5} q_H^{0.67} \tag{17}$$

Then, the suppression factor S can be modelled by a least square program. After various test, the final form of factor S is determined as the function of the convection and confinement number. The formula is expressed as follows:

$$S = 0.238 \frac{C_o^{0.238}}{C_f^{1.11}} \tag{18}$$

where the confinement number is defined as follows:

$$C_f = \frac{1}{D} \sqrt{\frac{\sigma}{g(\rho_l - \rho_g)}} \tag{19}$$

The proposed heat transfer coefficient correlation archives the mean deviation of 20.66 and 21.06% for macro-scale- and mini-scale channel, respectively. The detail comparison of the proposed correlation and the experimental data is shown in **Figure 12. Figure 13** summarises the overall comparison of existing heat transfer coefficient correlation and the proposed one with the present data of R410A. Among the existing correlations, the one proposed by Gungor-Winterton [31] showed the best prediction for the macro-channel with the mean deviation of 25% while this correlation failed to predict the heat transfer coefficient of mini-channel. The heat transfer coefficient correlation proposed by Tran et al. [33] predicted the experimental data with the mean deviation of around 30% for both macro- and mini-channel.

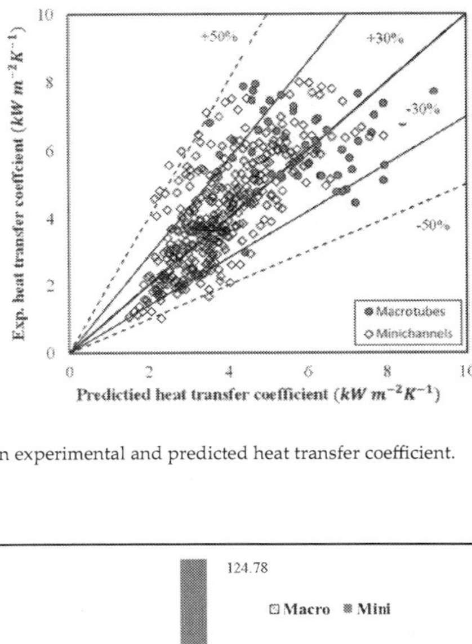

Figure 12. Comparison between experimental and predicted heat transfer coefficient.

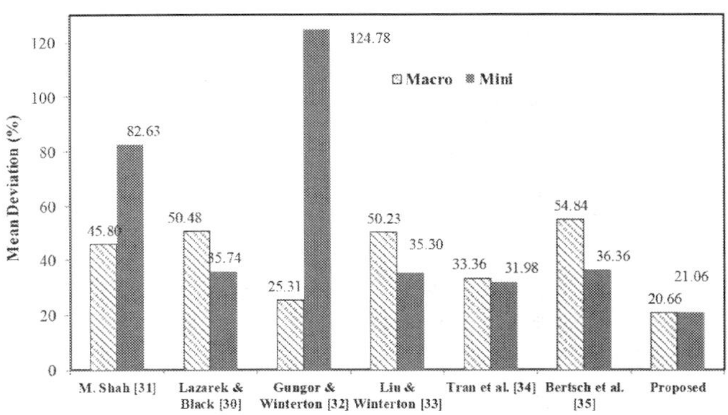

Figure 13. Summary of mean deviation.

4. Conclusions

The boiling heat transfer coefficient and pressure drop characteristics of R410A in horizontal tube with the inner diameter of 1.5, 3.0, 6.61 and 7.49 mm were demonstrated in this chapter. The data of heat fluxes ranging from 10 to 40 kW/m^2, mass fluxes ranging from 100 to 600 kgm^2/s, the vapour quality up to 1.0 and the saturation temperatures of 5–15°C. The results can be summarised as follows:

- The pressure drop of R410A is strongly affected by the mass flux, vapour quality and the inner tube diameter but the heat flux.
- The boiling heat transfer coefficient of R410A increased with the increase in mass flux, heat flux and the decreasing of inner tube diameter. However, in mini-channel, the effect of mass flux is only observed at the moderate quality regime, which indicates the predominance of nucleate boiling mechanism in mini-channel.
- The experimental pressure drop data were validated with some well-known correlations. Among them, the one developed by Gronnerud [20] showed the best prediction. A modified frictional pressure drop was proposed with the mean deviation of 9.29% versus the present data.
- The two-phase boiling heat transfer coefficient data were also compared with various general correlations. A new heat transfer coefficient correlation was also proposed that archived the good deviation versus the experimental data.

Nomenclature

A: Area (m^2)

AD: Average deviation, $\mathrm{AD} = \left(\frac{1}{n}\right)\sum_{1}^{n}\left(\left(\mathrm{d}p_{\mathrm{pred}} - \mathrm{d}p_{\mathrm{exp}}\right)\times 100 / \mathrm{d}p_{\mathrm{exp}}\right)$

Bo: Boiling number, $\mathrm{Bo} = \frac{q}{Gi_{\mathrm{fg}}}$

C: Chisholm parameter

c_p: Specific heat (kJ kg^{-1} K^{-1})

D: Diameter (m)

f: Friction factor

G: Mass flux (kg m^{-2} s^{-1})

g: Acceleration due to gravity (m s^{-2})

h: Heat transfer coefficient (kW m^{-2} K^{-1})

i: Enthalpy (kJ kg^{-1})

L: Tube Length (m)

MD: Mean deviation, $\mathrm{MD} = \left(\frac{1}{n}\right)\sum_{1}^{n}\left|\left(\left(\mathrm{d}p_{\mathrm{pred}} - \mathrm{d}p_{\mathrm{exp}}\right)\times 100 / \mathrm{d}p_{\mathrm{exp}}\right)\right|$

M: *Molecular weight* (kg kmol^{-1})

n: Number of data

p: Pressure (kPa)

Q: Electric power (kW)

q: Heat flux (kW m^{-2})

Re: Reynolds number, $\mathrm{Re} = \frac{GD}{\mu}$

RMS: Root mean square

T: Temperature (K)

$\dot{m}$: Mass flow rate (kg s^{-1})

X: Lockhart-Martinelli parameter

x: Vapour quality

z: Length (m)

Greek letters

α: Void fraction

Δi: The enthalpy rise across the tube (kJ kg^{-1})

μ: Dynamic viscosity (N s m^{-2})

ρ: Density (kg m^{-3})

σ: Surface tension (N m^{-1})

ϕ^2: Two-phase frictional multiplier

Gradients and differences

(dp/dz): Pressure gradient (N m^{-2} m^{-1})

(dp/dz F): Pressure gradient due to friction (N m^{-2}m^{-1})

http://dx.doi.org/10.5772/65966

Subscripts

crit: Critical point

exp: Experimental value

f: Saturated liquid

fi: Inlet liquid

g: Saturated vapour

i: Inner tube

lo: Liquid only

o: Outlet tube

pb: Pool boiling

pred: Prediction value

r: Reduced

sat: Saturation

sc: Subcooled

t: Turbulent

tp: Two-phase

v: Laminar

w: Wall

wi: Inside tube wall

Acknowledgement

This research was supported by Basic Science Research Program through the National Research Foundation of Korea (NRF) funded by Ministry of Education, Science and Technology (NRF-2016R1D1A1A09919697).

Author details

Jong-Taek Oh*[1], Nguyen Ba Chien[2], Kwang-Il Choi[1] and Pham Quang Vu[2]

*Address all correspondence to: ohjt@chonnam.ac.kr

1 Department of Refrigeration and Air-Conditioning, Chonnam National University, Yeosu, Chonnam, South Korea

2 Graduate School, Chonnam National University, Yeosu, Chonnam, South Korea

References

[1] Yongchan, K., Kookjeong, S., Jin, T. C. (2002). Evaporation heat transfer characteristics of R-410A in 7 and 9.52 mm smooth/micro-fine tubes. Int. J. Refrigeration, 25, 716–730.

[2] Man-Hoe, K., Joeng-Seob, S. (2005). Evaporating heat transfer of R22 and R410A in horizontal smooth and micro tubes. Int. J. Refrigeration, 28, 940–948.

[3] Wellsandt, S., Vamling, L. (2005). Evaporation of R407C and R410A in a horizontal herringbone microfin tube: heat transfer and pressure drop. Int. J. Refrigeration, 28, 901–911.

[4] Park, C. Y., Hrnjak, P. S. (2007). CO2 and R410A flow boiling heat transfer, pressure drop, and flow pattern at low temperatures in a horizontal smooth tube. Int. J. Refrigeration, 30(1), 166–178.

[5] Padovan, A., Del Col, D., Rossetto, L. (2011). Experimental study on flow boiling of R134a and R410A in a horizontal microfin tube at high saturation temperatures. Appl. Therm. Eng., 31(17–18), 3814–3826.

[6] Cavallini, A., Del Col, D., Doretti, L., Matkovic, M., Rossetto, L., Zilio, C. (2005). Two-phase frictional pressure gradient of R236ea, R134a and R410A inside multi-port mini-channels. Exp. Therm. Fluid Sci., 29(7), 861–870.

[7] Kaew-On, J., Wongwises, S. (2009). Experimental investigation of evaporation heat transfer coefficient and pressure drop of R-410A in a multiport mini-channel. Int. J. Refrigeration, 32(1), 124–137.

[8] C. Nguyen, V. Pham, Choi, K-I, OH, J-T, Heat Transfer Coefficient and Pressure Drop of R410A During Evaporation Inside Aluminum Multiport Minichannels, in: Vol. 2 Photovoltaics; Renewable-Non-Renewable Hybrid Power Syst. Smart Grid, Micro-Grid Concepts; Energy Storage; Sol. Chem. Sol. Heat. Cool. Sustain. Cities Communities, Transp. Symp. Integr. Buil, ASME, 2015: p. V002T18A004

[9] Choi, K.-I., Pamitran, A. S., Oh, C.-Y., Oh, J.-T. (2008). Two-phase pressure drop of R-410A in horizontal smooth minichannels. Int. J. Refrigeration, 31(1), 119–129.

[10] Oh, J. T., Pamitran, A. S., Choi, K. I., Hrnjak, P. (2011). Experimental investigation on two-phase flow boiling heat transfer of five refrigerants in horizontal small tubes of 0.5, 1.5 and 3.0 mm inner diameters. Int. J. Heat Mass Transf., 54(9–10), 2080–2088.

[11] Chien, N. B., Vu, P. Q., Choi, K.-I., Oh, J.-T. (2015). A general correlation to predict the flow boiling heat transfer of R410A in macro-/mini-channels. Sci. Technol. Built Environ., 21(5), 526–534.

[12] Lemmon, E. W., Huber, M. L., McLinden, M. O. (2007). NIST standard reference database 23: reference fluid thermodynamic and transport properties-REFPROP, Version 8.0. National Institute of Standards and Technology, USA.

[13] Steiner, D. (1993). Heat transfer to boiling saturated liquids. VDI Heat Atlas 1st edition. Society Process Engineering and Chemical Engineering. Springer-Verlag Berlin Heidelberg. (Translator: J. W. Fullarton, Dusseldorf.)

[14] Rouhani, Z., Axelsson, E. (1970). Calculation of volume void fraction in subcooled and quality region. Int. J. Heat Mass Transf., 13, 383–393.

[15] Choi, K.-I., Pamitran, A. S., Oh, J.-T., Saito, K. (2009). Pressure drop and heat transfer during two-phase flow vaporization of propane in horizontal smooth minichannels. Int. J. Refrigeration, 32(5), 837–845.

[16] Zhao, Y., Molki, M., Ohadi, M. M., Dessiatoun, S. V. (2000). Flow boiling of CO2 in microchannels. ASHRAE Trans. 106(Part 1), 437–445.

[17] Yoon, S. H., Cho, E. S., Hwang, Y. W., Kim, M. S., Min, K., Kim, Y. (2004). Characteristics of evaporative heat transfer and pressure drop of carbon dioxide and correlation development. Int. J. Refrigeration, 27, 111–119.

[18] Oh, H. K., Ku, H. G., Roh, G. S., Son, C. H., Park, S. J. (2008). Flow boiling heat transfer characteristics of carbon dioxide in a horizontal tube. Appl. Therm. Eng., 28, 1022–1030.

[19] Cho, J. M., Kim, M. S. (2007). Experimental studies on the evaporative heat transfer and pressure drop of CO2 in smooth and micro-fin tubes of the diameters of 5 and 9.52 mm. Int. J. Refrigeration, 30, 986–994.

[20] Gronnerud, R. (1972). Investigation in liquid holdup, flow resistance and heat transfer in circular type evaporator, part IV: two-phase resistance in boiling refrigerant. Bulletin de l'Inst du Froid, Annexe, 1972.

[21] Chisholm, D. (1973). Pressure gradient due to the friction during the flow of evaporating two-phase mixture in smooth tubes and channel. Int. J. Heat Mass Transf., 16, 347–358.

[22] Friedel, L. (1979). Improve friction pressure drop correlation for horizontal and vertical two-phase pipe flow. European Two-Phase Flow Group Meeting, Ispra, Italia, June, Paper E2.

[23] Muller-Steinhagen, H., Heck, K. (1986). A simple friction pressure drop correlation for two phase flow in pipe. Chem. End. Process., 20, 297–308.

[24] Lockhart, R. W., Martinelli, R. C. (1949). Proposed correlation of data for isothermal two-phase, two-component flow in pipes. Chem. Eng. Prog., 45, 39–48.

[25] Kew, P. A., Cornwell, K. (1997). Correlations for the prediction of boiling heat transfer in small-diameter channels. Appl. Therm. Eng., 17(8–10), 705–715.

[26] Lazarek, G. M., Black, S. H. (1982). Evaporative heat transfer, pressure drop and critical heat flux in a small diameter vertical tube with R-113. Int. J. Heat Mass Transf., 25, 945–960.

[27] Wambsganss, M. W., France, D. M., Jendrzejczyk, J. A., Tran, T. N. (1993). Boiling heat transfer in a horizontal small-diameter tube. J. Heat Transf., 115, 963–975.

[28] Tran, T. N., Wambsganss, M. W., France, D. M. (1996). Small circular-and rectangular-channel boiling with two refrigerants. Int. J. Multiphase Flow, 22(3), 485–498.

[29] Mastrullo, R., Mauro, A. W., Thome, J. R., Toto, D., Vanoli, G. P. (2012). Flow pattern maps for convective boiling of CO2 and R410A in a horizontal smooth tube: experiments and new correlations analyzing the effect of the reduced pressure. Int. J. Heat Mass Transf., 55(5–6), 1519–1528.

[30] Shah, M. M. (1982). Chart correlation for saturated boiling heat transfer: equations and further study. ASHRAE Trans. (United States), 88(1), 185–196.

[31] Gungor, K., Winterton, R. (1986). A general correlation for flow boiling in tubes and annuli. Int. J. Heat Mass Transf., 29(3), 351–358.

[32] Liu, Z., Winterton, R. (1991). A general correlation for saturated and subcooled flow boiling in tubes and annuli, based on a nucleate pool boiling equation. Int. J. Heat Mass Transf., 34(11), 2759–2766.

[33] Tran, T., Wambsganss, M., France, D. (1996). Small circular-and rectangular-channel boiling with two refrigerants. Int. J. Multiphase Flow, 3(4), 485–498.

[34] Bertsch, S. S., Groll, E. A., Garimella, S. V. (2009). A composite heat transfer correlation for saturated flow boiling in small channels. Int. J Heat Mass Transf., 52(7–8), 2110–2118.

[35] Kandlikar, S. (1990). A general correlation for saturated two-phase flow boiling heat transfer inside horizontal and vertical tubes. J. Heat Transf., 112(1), pp. 219–228.

[36] Kandlikar, S., Steinke, M. (2003). Predicting heat transfer during flow boiling in minichannels and microchannels, ASHRAE Trans., 109, 667–676.

[37] Chen, J. C. (1966). A correlation for boiling heat transfer to saturated fluids in convective flow. Ind. Eng. Chem. Process Des. Dev., 5, 322–329.

[38] Dittus, F. W., Boelter, L. M. K. (1930). Heat transfer in automobile radiators of the turbulent type. Univ. Calif. Publ. Eng., 2, 443–461.

[39] Cooper, M. G. (1984). Heat flow rates in saturated nucleate pool boiling e a wide-ranging examination using reduced properties. Adv. Heat Transf., 16, 157–239.

Chapter 3

Comprehensive Study of Compact Heat Exchangers with Offset Strip Fin

Latife Berrin Erbay, Mehmet Mete Öztürk and Bahadır Doğan

Additional information is available at the end of the chapter

http //dx.doi.org/10.5772/66749

Abstract

This chapter is aimed to address the performance of compact heat exchangers with offset strip fin, which have been studied by the researchers in detail for decades. The history and basic features of offset strip fins (OSF) are described first to introduce the fin geometry. Then, the effect of the fin geometry on the performance of the offset strip fin is given from experimental and numerical aspects, respectively. Flow streams evolution under varying offset strip geometries is summarized in order to demonstrate the physical impact on the flow. The thermohydraulic features of the flow in the offset strip fin are investigated by considering the Colburn *j*-factor and friction (*f*) factors in diverse flow regimes. Furthermore the criteria, flow area goodness factor j/f, the ratio $j/f^{1/3}$ and thermohydraulic performance factor *JF*, derived from the mentioned dimensionless factors, are also used as a scale of the performance of the structure and reported in the chapter.

Keywords: heat exchanger, offset strip fin

1. Introduction

The heat exchangers are one of the crucial components of machines since to remove the heat from the running system is the major concern, in order to improve the proper functioning. Numerous efforts have been performed by the researchers to overcome this issue by several methods and systems. The typical solution, which has been in use for decades, is the usage of heat exchangers, which are developed depending on the particular demands of the systems and have met the needs and improved the system's performance. But as in the other equipment, it came to a point where it is not sufficient enough to overcome the requirements. The compact heat exchangers emerged almost six decades back based on these particular demands. The most common definition of the compact heat exchangers in literature is "a heat exchanger having a surface area density greater than about $700 m^2/m^3$" [1]. The ultimate purpose of the

INTECH
open science | open minds

© 2017 The Author(s). Licensee InTech. This chapter is distributed under the terms of the Creative Commons Attribution License (http://creativecommons.org/licenses/by/3.0), which permits unrestricted use, distribution, and reproduction in any medium, provided the original work is properly cited.

studies about the compact heat exchangers is to produce more efficient ones by reducing the physical sizes of the equipment for a given duty, which leads to use less coolant as well. There is not much possibility in order to get this goal, but one of these options is to have a higher heat transfer rate for particular conditions and the other one is to create a higher surface area and the last one is increasing both. The typical way to increase heat transfer surface area is using fins on the heat exchangers, which provide a higher surface area per unit volume ratio. The researchers endeavor to develop more efficient heat exchangers but small passage dimensions, nonuniformities and geometrical changes make it hard to characterize the heat transfer surface. The applications of the compact heat exchangers can be widely found in industry such as air conditioning, refrigeration, automotive and aerospace.

In this particular chapter, it is aimed to inform and address the offset strip fins, which have been studied by the researchers in detail for decades and still getting the attention due to its superior advantages. In the following sections, the history and fundamentals of this structure will be given first and in the following parts, the investigations will be summarized by considering their objectives, which are handled in the communications such as parametric effect of the structure, experimental and numerical research of the fin under varying flow regimes and conditions and the evolution of heat transfer and friction factors under different flow conditions by the change of the regime. The chapter will be concluded with the remarks that will outline the findings and will guide to the future studies.

2. History and basics of offset strip fin

Offset strip fin is one of the most preferred fin geometry in compact heat exchangers, which has a rectangular cross section cut into small strips of length, l and displaced by about 50% of the fin pitch in the transverse direction. The scheme of a typical strip fin is presented in **Figure 1**. The most significant variables of fin geometry are the fin thickness and strip length in the flow direction that leads to higher heat transfer coefficient and higher friction factors than plain fin geometries. The main reason of this improvement relies on developing laminar boundary layers (**Figure 2**). The enhancement is provided by the interruption of the flow

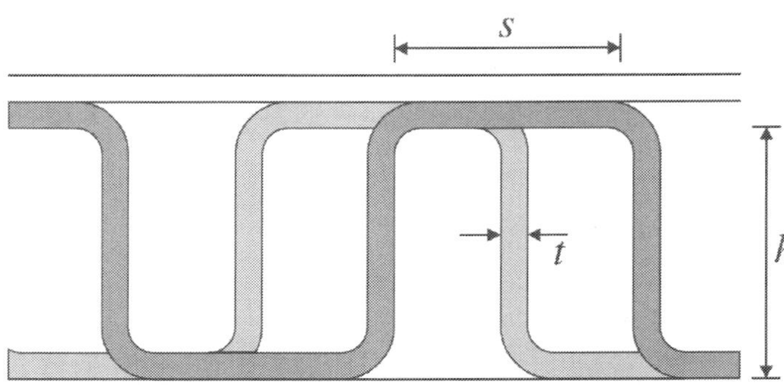

Figure 1. Offset strip fin schematic representation [2].

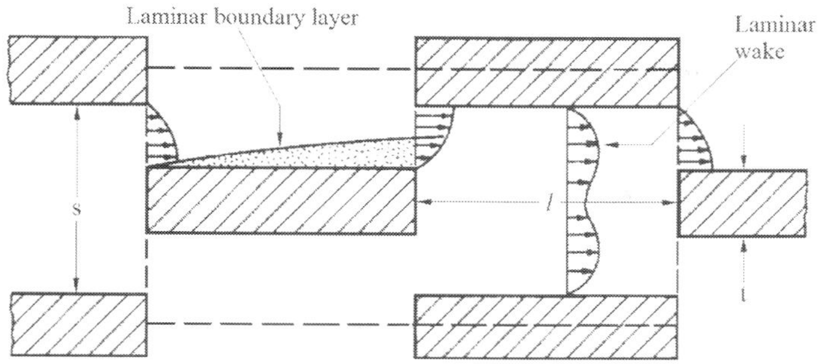

Figure 2. Flow patterns observed in the visualization experiments [3].

periodically, which does not only create fresh boundary layers but also generate greater viscous pressure drop due to higher friction factor (**Figure 2**).

The surface geometry in the given representation is described as follows: the fin length is *l*, height is *h*, transverse spacing is *s* and thickness is *t*. Even though nonuniform spacing is applicable and possible, generally the fin offset is the half-fin spacing and uniform. Furthermore, there are some other parameters that have influence on the flow and heat transfer like manufacturing irregularities (such as burred edges, bonding imperfections, etc.) [4].

Even though offset strip fins have been studied for decades by numerous researchers, Kays and London [5] and London and Shah [6] could be easily called as the pioneers of the offset strip fin researches by their valuable reports of their experiments about offset strip fins. The roots of their investigations rely on a test program at Stanford University in 1947 [5]. Since then, they have published their outcomes in several reports and papers, which still keep their importance to enlighten the path of researchers working in this field. In their study, they have examined several types of fins with regard to their varying parameters and operating conditions in order to explain and uncover blurry parts of the compact heat exchangers. In one of their very valuable publications [6], they shared their outcomes on the offset strip fins. The schematic representation of the fin structure is presented in **Figure 3**.

In that particular study, they have examined eight different fins, which differ from each by their fin numbers per inch, type of flow section, fin height, distance between plates, flow length, fin thickness and the material used in the study. Due to diverse fin type used in their studies, Kays and London [5] and London and Shah [6] have developed a coding method (explanation of the code is given in **Table 1**) in order to differentiate the fins from each other. An example of that coding might be given as in the following,

$$\underset{1}{25}.\underset{2}{01}.\underset{3}{R(S)} - \underset{4}{0.201}/\underset{5}{0.200} - \underset{6}{1/9}\underset{7}{(O)} - \underset{8}{0.004}\underset{9}{(A1)}$$

The numbers given underneath the code stands for, number of fins per inch, type of fin flow cross section (R=rectangular, T=triangular, U=U shape), fin sandwich construction

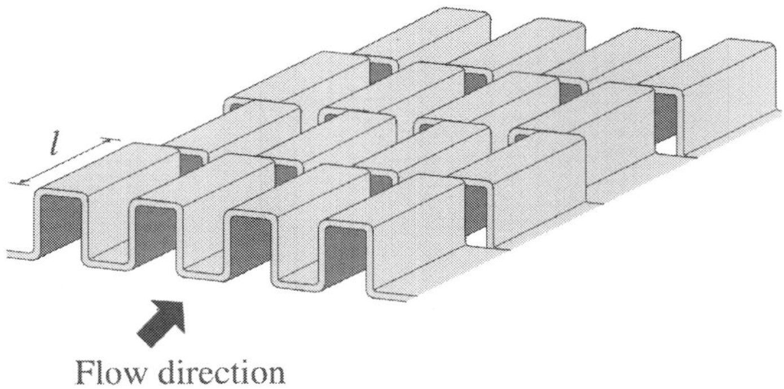

Figure 3. General features of the fin core material [6].

No	Explanation of the code
1	Number of fins per inch
2	Type of fin flow cross section
3	Fin sandwich construction
4	Fin height before brazing (inch)
5	Fin height after brazing (inch)
6	Uninterrupted fin length (inch)
7	Type of surface
8	Fin metal thickness (inch)
9	Fin material

Table 1. Explanation of codes.

(SD=single-double, S=single, D=double, T=triple), fin height before brazing, fin height after brazing, uninterrupted fin length, type of surface (L=louvered, O=offset, P=plain, S =strip), fin metal thickness, fin material (Al=aluminium, SS=stainless steel, Ni=nickel, etc.) respectively. Since their experiments are the basics of the OSF heat exchanger researches, it would be better to understand their terminology to distinct the varying geometries and structures in their investigations.

3. Data reduction of offset strip fins

The aim of the thermohydraulic analysis of the offset strip fins is to determine the pressure drop data and overall heat transfer coefficients of the structure. The pressure drop could be directly obtained from experiments, while the heat transfer rate could be found from experimental measurements by applying energy balance equations on either hot or cold streams. The typical equations used in data reduction could be derived as in the following.

In order to determine the heat transfer rate, the effectiveness-NTU method is used, which would be the ultimate purpose of the analysis.

The average heat transfer rate can be calculated by using Eq. (1)

$$\dot{Q} = (\dot{Q}_c + \dot{Q}_h)/2 \quad (1)$$

where $\dot{Q}_c$ and $\dot{Q}_h$ are the heat transfer rates of cold and hot stream, respectively. The heat transfer rates of each fluid can be calculated with Eq. (2) and Eq. (3)

$$\dot{Q}_c = \dot{m}_c c_{p,c}(T_{c,o} - T_{c,i}) \quad (2)$$

$$\dot{Q}_h = \dot{m}_h c_{p,h}(T_{h,i} - T_{h,o}) \quad (3)$$

The effectiveness for both unmixed fluid,

$$\varepsilon = 1 - exp\left[\frac{NTU^{0.22}}{C_r}\{exp(-C_r NTU^{0.78}) - 1\}\right] \quad (4)$$

where

$$\varepsilon = \dot{Q}/\dot{Q}_{max} \quad (5)$$

$$C_r = \frac{(\dot{m}c_p)_{min}}{(\dot{m}c_p)_{max}} \quad (6)$$

With regard to these, the overall heat transfer coefficient (UA) can be obtained for the heat exchanger as,

$$UA = (\dot{m}c_p)_{min} NTU \quad (7)$$

where

$$UA = \dot{Q}/\Delta T_m \quad (8)$$

ΔT_m is the logarithmic mean temperature difference, which can be defined as,

$$\Delta T_m = \frac{(T_{h,i} - T_{c,o}) - (T_{h,o} - T_{c,i})}{\ln\left((T_{h,i} - T_{c,o})/(T_{h,o} - T_{c,i})\right)} \quad (9)$$

The heat transfer coefficient (h_c) for the cold stream can be written as in the following,

$$\frac{1}{\eta_c A_c h_c} = \frac{1}{UA} - \frac{\delta_t}{k_t A_t} - \frac{1}{A_h h_h} \quad (10)$$

In this equation, h_h is the heat transfer coefficient of the hot stream. One of the other parameters in the given equation is the surface effectiveness (η_c) for a dry surface,

$$\eta_c = 1 - \frac{A_f}{A_c}(1 - \eta_f) \tag{11}$$

where η_f is,

$$\eta_f = \frac{\tanh(ml)}{ml} \tag{12}$$

where

$$m = \sqrt{h_c P_f / k A_{c,f}} \tag{13}$$

The flow regime is the crucial parameter of the thermohydraulic analysis of the heat exchanger studies since it has a big impact on the performance of the fin structure. Even though Reynolds number has a unique definition, it may be interpreted by the researchers from various aspects and could be written in different ways with regard to defined hydraulic diameter,

$$Re = \frac{u\, D_e}{\upsilon} \tag{14}$$

where hydraulic diameter D_e is,

$$D_e = \frac{2sh}{s + h} \tag{15}$$

As noted before, the major difference between the *Re* number relies on the definition of the hydraulic diameter and in order to provide a better look, some of the reported diameters and the corresponding *Re* numbers will be summarized in the following **Table 2**.

Reference	Hydraulic diameter	Reynolds number
[4]	$D_e = \frac{2 \bullet (s-t)h}{(s+t)+ht/l}$	$Re = \frac{\rho \upsilon D_e}{\mu}$
[7]	$D_e = \frac{2 \bullet (H_f - t_f)(P_f - t_f)}{(H_f - t_f)+(P_f - t_f)}$	$Re = \frac{u\, D_e}{\upsilon}$
[8]	$D_e = \frac{4 \bullet s \bullet h_f \bullet l}{2 \bullet (s \bullet l + h_f \bullet l + h_f \bullet t) + s \bullet t}$	
[9]	$D_e = \frac{2 \bullet w \bullet L(F_p - t)}{w \bullet (L+t) + L \bullet (F_p - t)}$	
[10]	$D_e = \frac{4 \bullet s \bullet h \bullet l}{2 \bullet (s \bullet l + h \bullet l + h \bullet t) + s \bullet t}$	$Re_{D_e} = \frac{\rho u D_e}{\mu}$
[11]	$D_e = \frac{2 \bullet (s-b)l}{(l+b)}$	$Re = \frac{u_c D_e}{\upsilon}$

Table 2. Hydraulic diameter described in offset strip fin investigations from literature.

3.1. Heat transfer and pressure drop characteristics of offset strip fins

Like all other systems and devices, there are some nondimensional factors to evaluate the performance of the offset strip fins. The most common parameters used to decide the benefits of the structure are Colburn *j*-factor and friction factor (*f*), which corresponds to the heat transfer and pressure drop, respectively. These two parameters could be defined as in the following,

$$j = StPr^{2/3} \tag{16}$$

$$f = \left(\frac{A_c}{A}\right)\left(\frac{2\Delta P}{\rho u^2} - k_c - k_e\right) \tag{17}$$

where A_c is the minimum free flow area for the external side and k_c and k_e are the coefficients of pressure loss at the inlet and the outlet of the heat exchanger. In addition to this, the Stanton number and the Prandtl number, which are used to define the Colburn j-factor, are,

$$St = \frac{h_c}{\rho u c_p} \tag{18}$$

$$Pr = \frac{\mu c_p}{\lambda} \tag{19}$$

where h_c is heat transfer coefficient, ρ is density, u velocity of the fluid, c_p specific heat of the fluid, μ dynamic viscosity and λ thermal conductivity. Since these two parameters are commonly preferred by the researchers in order to observe the performance of the offset strip fins, the results received at the studies with regard to these two parameters will be summarized in the coming sections where the published papers will be addressed.

One of the first friction factor and Colburn factor investigations was performed by London and Shah [6] as they were the leading investigators in this field. In the study, the effect of blockages either in steam side (the side where the steam is introduced) or in the air side (the side where the air flows across the offset fin) is underlined. The reason for the first one is explained as the poor brazing, which tends to condensation and bridge flow passage, whereas the latter one is because of bent fin edges and results in lower Colburn j-factor and higher friction factor, f. Furthermore, the effect of nondimensional parameters on j and f for different cores (different fin structures with different fin numbers) is highlighted as well. It is worth to note that, with regard to their findings when the aspect ratio gets lower, the friction factor and heat transfer are affected less and while the number of fins per unit size gets higher, they are affected more. In addition, offset spacing length is illustrated in **Figure 4**. The importance of the effect is emphasized in **Figure 4**, where increasing the fin spacing lowers the friction factor and heat transfer and vice versa and lowering the spacing makes the performance of the fin structure higher.

In addition to these most common performance criterions, new approaches are also suggested by the researchers. Bhowmik and Lee [7] adopted the j/f, $j/f^{1/3}$ and JF in order to examine the performance of the offset strip fins instead of conventional methods. j/f is known as "area goodness factor" [7] and $j/f^{1/3}$ is known as "volume goodness factor" [7]. JF number, which is related with the volume goodness factor, can be obtained by the following equation [7].

$$JF = \frac{j/j_R}{(f/f_R)^{1/3}} \tag{20}$$

where j_R and f_R are the reference values of Colburn j-factor and friction factor, respectively.

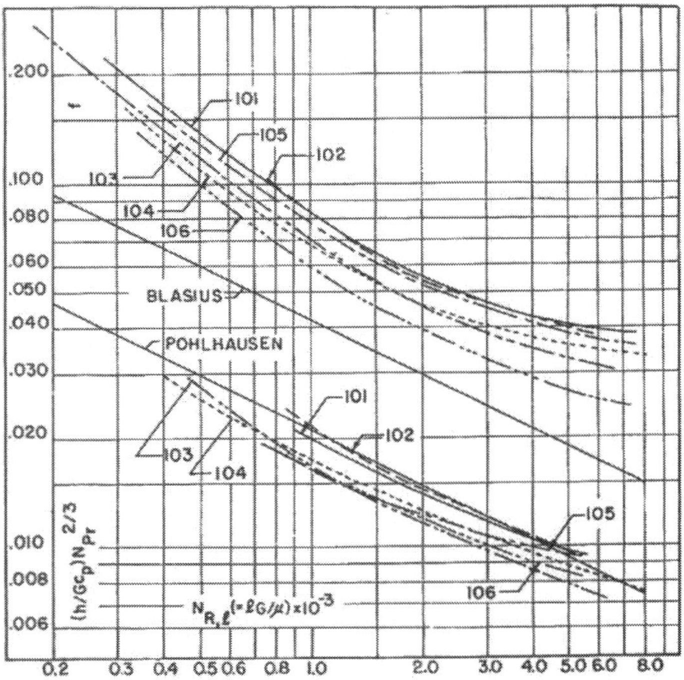

Figure 4. Offset spacing length vs. Reynolds number comparisons for varying surfaces [6].

The analyses are performed by a commercial computerized fluid dynamics (CFD) software in three dimensions. Thermohydraulic performance is studied as well for laminar, transition and turbulent flow regions, which lead to a general correlation that fits to all three regions. Since varying fluids are investigated as well, a *Nu* correlation including the Prandtl effect is also presented. Firstly, the numerical computations are validated by Manglik and Bergles [4] and Joshi and Webb [3] correlations and seen how they have a good agreement with them and due to their continuity, the results are claimed as a good candidate for a single correlation that covers all three flow regions.

Unlike the earlier reports, the performance of the offset strip fins are examined by flow area goodness factor *j/f*, the ratio $j/f^{1/3}$ and thermohydraulic performance factor *JF*. In these performance characteristics, a high *j/f* factor provides a low frontal area for the heat exchangers. When the *JF* factor is high, it refers to a good thermohydraulic performance and finally, high $j/f^{1/3}$ factor leads to a good heat transfer and pressure loss performance.

The effect of these three factors is observed for five Prandtl numbers (*Pr* =0.7, 7, 16, 33 and 50), which will enable to decide which are the right working fluid conditions as shown in **Figure 5**. The *Pr* effect is more significant for *j/f* factor at the turbulent region; $j/f^{1/3}$ factor increases when

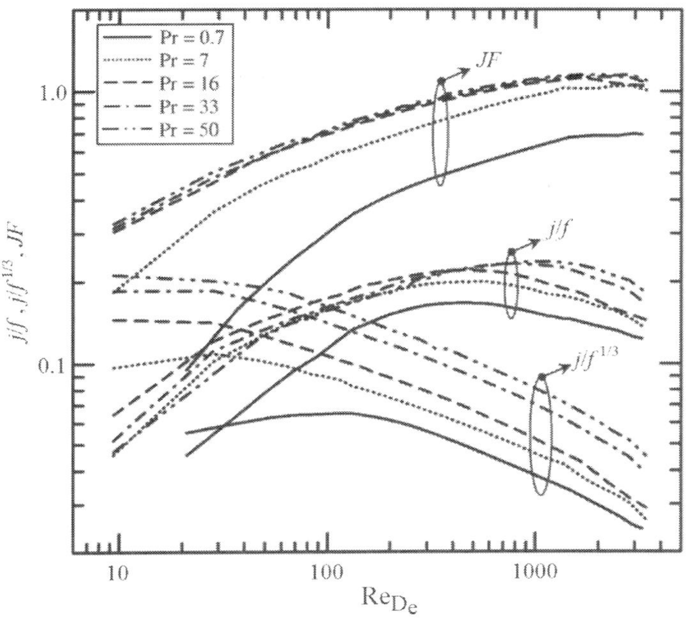

Figure 5. Evolution of performance factors [7].

the *Pr* ascends on the other side; unlike the other two factors, *Pr* does not have a distinctive effect on *JF*. According to these, it is underlined that *JF* factor could be useful for the water, while $j/f^{1/3}$ is more convenient for gas-oil liquid. In contrary to the given two factors, *j/f* could not be considered as a good performance criterion for fluids.

Moreover, these performance criteria are employed to provide a better comparison between seven common configurations namely plain, perforated, offset strip, louvered, wavy, vortex-generator and pin of plate-fin heat exchangers by Khoshvaght-Aliabadia et al. [8]. Water is used as the working fluid in Reynolds number range between 480 and 3770. The other purpose is to select an optimum plate-fin channel in which the best performance criteria evolution is observed. The significant enhancement is observed at the vortex-generator channel in the heat transfer coefficient and a proper reduction in the heat exchanger surface area (**Figure 6**). Moreover, the wavy channel displays an optimal performance at low Reynolds numbers, according to the referred criteria's. The offset strip fins, *f* factor, ascends by the increasing *Re* number as predicted but it is noted that the critical value for *Re* is 1800 and beyond that the friction factor rises 11.7%. As for the $j/f^{1/3}$, the offset strip fins got the highest values when the *Re* >1500, while wavy has the highest for the low *Re* range. Furthermore, it is observed that the *JF* factor curves show the larger increment by ascending *Re* number for offset strip and vortex generators.

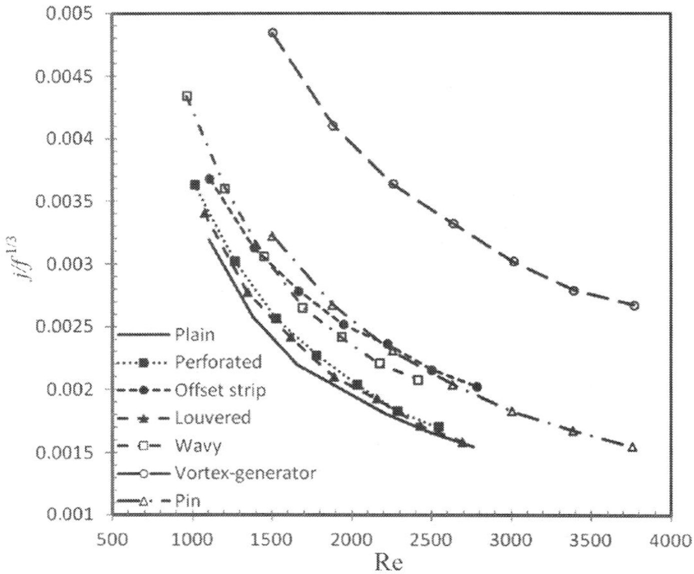

Figure 6. $j/f^{1/3}$ ratio vs. *Re* for different plate channels [8].

A different performance evaluation aspect is employed to the OSF heat exchangers by considering second law of thermodynamics [9]. A new parameter that is called as relative entropy generation distribution factor is proposed by the researchers. This new parameter represents a ratio of relative changes of entropy generation. The effect of parameters, which are commonly used nondimensional parameters for the OSF studies, is discussed. Optimization for the investigated parameters is carried out, which could provide sufficient information about the conditions that could maximize the performance. The proposed performance evaluation parameter bases on the entropy generation number (N_{s1}) and can be written as in the following,

$$\psi^* = \frac{(N_{s1,o,\Delta T} - N_{s1,a,\Delta T})/N_{s1,o,\Delta T}}{(N_{s1,a,\Delta P} - N_{s1,o,\Delta P})/N_{s1,o,\Delta P}} = \frac{1 - N^*_{s1,\Delta T}}{N^*_{s1,\Delta P} - 1}. \tag{21}$$

where 'a' and 'o' refer to the augmented (OSF) and reference (plain plate-fin) channel, respectively and 'ΔT' and 'ΔP' stand for entropy generation due to the temperature difference and pressure drop, respectively.

The given performance is examined for different geometrical parameters in order to estimate the effective ones, by considering varying relative temperature difference. According to the obtained data, smaller thickness at lower flow rates gives better results.

4. Experimental studies for influence on flow and heat transfer

One of the fundamental research topics about offset strip fin is its geometry. The effect of the geometry on the performance of the fin and heat exchanger is the main concern of the investigators for decades. For this purpose, some researchers kept working on the effect of parameters on the performance of offset strip fin experimentally along the same path developed [5, 6]. One of these studies was released by Dong et al. [10]. Experiments are carried on air side of the heat exchanger to figure out the evolution of j and f for 16 different offset strip fins by flat tube heat exchanger. The flow of the air is ranging in $500<Re<7500$, while the water flow is kept constant at $2.5 m^3/h$. The effect of fin geometry is presented for: space, height, length and flow length. In addition to the conventional approach to the fin structure, few researchers try to implement a new aspect to the studies. One of these investigations was reported by Peng et al. [11], which discusses the performance of a compact heat exchanger by an innovative offset strip fin array (**Figure 7**) at a range Re =500–5000. With regard to the results, when fin pitch increases, the heat transfer reduces while the friction factor reaches to higher values. In addition, the fin length effect is presented as well, with regard to that heat transfer will be enhanced by the shorter lengths, while the friction factor increases as well. One other valuable analysis that can be pointed out is the effect of fin bending; according to them, by the increment of fin bending, the heat transfer increases and friction factor reduces.

Plate fin heat exchangers are commonly used in the air to air or air to gas applications, but not so appropriate to the liquid to liquid applications, which also have crucial importance for some applications. In order to fulfill this requirement, brazed plate heat exchangers are emerged. Commercially available ones are made up of stainless steel that is brazed by copper or nickel. Even though they meet the higher pressure and temperature conditions, they are not sufficient for the applications where corrosion issues appeared. In order to overcome this deficit, Fernández-Seara et al. [12] studied titanium brazed plate fin heat exchangers with OSF. The corrosion resistance lets titanium to be used in liquid-liquid applications, unlike the typical fin

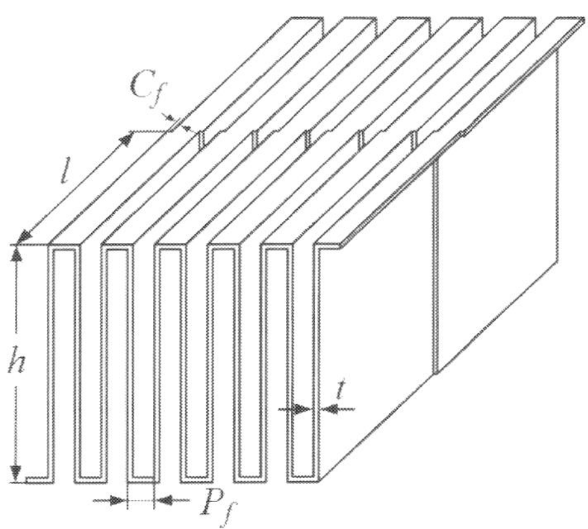

Figure 7. Scheme of the innovative offset strip fin [11].

structures. Across the heat exchanger, water-water and 10–30% ethylene glycol are used in the experiments. As a result of the investigations, empirical correlation to determine single phase convection heat transfer coefficient, as a function of *Re* is derived. The importance of this correlations are underlined since the experiments are carried out for water and ethylene glycol, which have higher numbers than the air and tends to provide more accurate predictions than the ones derived by air. The outcomes of the research are presented for varying *Re* number of water-water test and ethylene glycol mixture, individually. Pressure drop and overall heat transfer coefficients of the particular conditions are summarized with respect to the velocities or mass flow rate that they are tested. It is noted that the pressure drop gets higher as the ethylene glycol percentage increases when the flow regime drives up. In addition, the heat transfer is poor in ethylene glycol when it is compared with water and it gets worse when the mass flow reduces.

Offset strip fins are usually operated at low Reynolds numbers ($Re<1000$), at which the flows are typically steady and laminar. Researchers have investigated the thermohydraulic performances of the offset strip fins widely at this regime. Unlike the reported studies, Michna et al. [13] tried to find the answer for the thermohydraulic analysis of the OSF at turbulent regime. The friction factor and heat transfer coefficient of the OSF are measured at *Re* number in the range $1000<Re<10{,}000$ in the experiments. In order to have a better approach, pressure and heat transfer tests are executed by aluminium and stainless steel, respectively, where the constant heat flux is provided by film heaters. For the confirmation of the friction factor, the correlation of Manglik and Bergles [4] (the correlation given in **Table 3**) is used, even though the correlation is for $Re<20{,}000$, the correlation is extrapolated for the higher values of *Re*. An agreement with the correlation could be observed for friction factor at $Re<20{,}000$ and it starts to oscillate beyond that range, while Colburn *j*-factor consistently has higher values than the correlation of Manglik and Bergles [4]. The evaluation of the modified Colburn *j*-factor is presented row by row for varying *Re* values as well. With regard to the vortex shedding and turbulent flow, the highest values of j and f are observed for the higher values of Reynolds [13].

Even though the uniform distribution is expected to obtain higher performances in compact heat exchangers, maldistribution could also be seen, especially for parallel channel heat exchangers. There are limited studies published on this topic for offset strip fins. One of those few studies is released by Saad et al. [14], in which the experimental distribution of phases and pressure drop in offset strip fin type compact heat exchangers is given. The experiments are accomplished in a flat vertical compact heat exchanger in which the flow of air and water are examined. The experiments are carried out for single phase and two phase flows by order. The single phase flow experiments are followed by CFD analysis of the same problem employed in the investigation. Furthermore, a correlation (**Table 3**) that belongs to single phase flow is developed as well for prediction of friction factor. The nonuniformity of the two phase flow regime is determined by the maldistribution parameters such as standard deviation (STD) and normalized standard deviation (NSTD). The pressure drop profiles and flow rate distribution are used to identify the phase distribution of the flow since it is hard to determine the solely phase distribution without other physical features. The results of the experimental and CFD

Reference	Correlation
[2]	$f = 9.6243Re^{-0.7422}\alpha^{-0.1856}\delta^{0.3053}\gamma^{-0.2659}\times[1+7.669\times10^{-8}Re^{4.429}\alpha^{0.920}\delta^{3.767}]$ $j = 0.6522Re^{-0.5403}\alpha^{-0.1541}\delta^{0.1499}\gamma^{-0.0678}\times[1+5.269\times10^{-5}Re^{1.340}\alpha^{0.504}\delta^{0.456}\gamma^{-1.055}]^{0.1}$
[4]	Laminar range $f = 8.12Re_D^{-0.74}(l/D_e)^{-0.41}\alpha^{-0.02}$ $j = 0.53Re_D^{-0.5}(l/D_e)^{-0.15}\alpha^{-0.14}$ Turbulent range $f = 1.12Re_D^{-0.36}(l/D_e)^{-0.65}\alpha^{0.17}$ $j = 0.21Re_D^{-0.4}(l/D_e)^{-0.24}\alpha^{-0.02}$
[6]	$\beta < 20\%$ $f_{air} = exp(7.91)(\alpha)^{-0.159}(\delta)^{0.358}(\gamma)^{-0.033}(Re_{D_e})^{(0.126lnRe_{D_e}-2.3)}$ $j_{air} = 0.651(\alpha)^{-0.136}(\delta)^{0.236}(\gamma)^{-0.158}(Re_{D_e})^{(0.015lnRe_{D_e}-0.623)}$ $20\%\leq\beta < 25\%$ $f = exp(9.36)(\alpha)^{-0.0025}(\delta)^{-0.0373}(\gamma)^{1.85}(Re_{D_e})^{(0.142lnRe_{D_e}-2.39)}$ $j_{air} = 1.18(\alpha)^{-0.134}(\delta)^{0.0373}(\gamma)^{0.118}(Re_{D_e})^{(0.0445lnRe_{D_e}-0.982)}$ $25\%\leq\beta < 30\%$ $f = exp(5.58)(\alpha)^{-0.36}(\delta)^{0.552}(\gamma)^{-0.521}(Re_{D_e})^{(0.111lnRe_{D_e}-1.87)}$ $j_{air} = 0.49(\alpha)^{-0.23}(\delta)^{0.245}(\gamma)^{-0.733}(Re_{D_e})^{(0.049lnRe_{D_e}-0.971)}$ $30\%\leq\beta < 35\%$ $f = exp(4.84)(\alpha)^{-0.48}(\delta)^{0.347}(\gamma)^{0.511}(Re_{D_e})^{(0.089lnRe_{D_e}-1.49)}$ $j_{air} = 0.22(\alpha)^{-0.315}(\delta)^{0.235}(\gamma)^{-0.727}(Re_{D_e})^{(0.0313lnRe_{D_e}-0.729)}$ *By Prandtl effect* $\beta < 20\%$ $j_{Pr} = exp(1.96)(\alpha)^{-0.098}(\delta)^{0.235}(\gamma)^{-0.154}(Re_{D_e})^{(0.00643lnre_{D_e}-1.3)}(Pr)^{0.00348}$ $20\%\leq\beta < 25\%$ $j_{Pr} = 1.06(\alpha)^{-0.1}(\delta)^{0.131}(\gamma)^{-0.08}(Re_{D_e})^{(0.0323lnRe-0.856)}(Pr)^{0.0532}$ $25\%\leq\beta < 30\%$ $j_{Pr} = exp(1.3)(\alpha)^{0.004}(\delta)^{0.251}(\gamma)^{0.031}(Re_{D_e})^{(0.0507lnRe_{D_e}-1.07)}(Pr)^{0.051}$ $30\%\leq\beta < 35\%$ $j_{Pr} = 0.2(\alpha)^{-0.125}(\delta)^{0.21}(\gamma)^{-0.069}(Re_{D_e})^{(0.0005lnRe_{D_e}-0.338)}(Pr)^{0.0549}$
[12]	$10<Re<3500$ $f_{cor} = 10Re_{D_e}^{-0.68}$ $j_{cor} = 0.489Re_{D_e}^{-0.445}$
[10]	$f = 20.362Re_D^{-0.7661}$
[15]	Laminar range $f = 7.6616Re_D^{-0.712}(x/D)^{-0.384}\alpha^{-0.092}$ $j = 0.483Re_D^{-0.536}(x/D)^{-0.162}\alpha^{-0.184}$ Turbulent range $f = 1.136Re_D^{-0.198}(x/D)^{-0.781}(t/D)^{0.534}$ $j = 0.242Re_D^{-0.368}(x/D)^{-0.322}(t/D)^{0.089}$
[16]	$500<Re<7500$ $f = 2.092Re^{-0.281}\alpha^{-0.739}\beta^{0.972}\delta^{-0.78}\gamma^{-0.497}$ $j = 0.101Re^{-0.189}\alpha^{-0.488}\beta^{0.479}\delta^{-0.279}\gamma^{-0.315}$

Table 3. Heat transfer and friction factor correlations for offset strip fins.

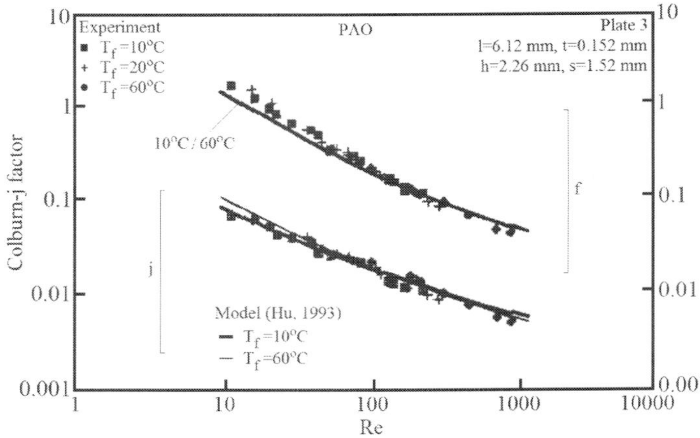

Figure 8. A comparison of test and numerical results of coolant polyalphaolefin [15].

simulation are confirmed by Manglik and Bergles [4] correlation (**Table 3**). A similar trend of that correlation, which is the descending of the friction factor by ascending *Re* number, is observed in the reported results. The correlated friction factor and the comparison of it with the reported ones are summarized and given in detail in correlations of *j* and *f* factors in OSF heat exchangers (HEX) section in order to give a better aspect to the findings.

Although the scope of the study is about OSF compact heat exchangers, which mostly deal with air, this is not the only fluid used with OSF. So, it is important to note the thermal hydraulic response of the fin structure with different *Pr* numbers. The effect of Prandtl is presented in a series of investigation by Hu and Herold, either experimentally [15] and numerically [16]. The experiments are corresponding to a liquid (water and polyalphaolefin) cooled offset fin compact heat exchanger. As for the given liquids, the *Pr* number that is used in the study ranges from 3 to 150. The obtained results compared by the air cooled models and a noticeable difference is observed for varying Prandtl values. A numerical analysis is performed to investigate uniformity of heat flux and temperature distribution and also for the validation of the tested conditions, where a good agreement could be seen (**Figure 8**). The model results were used to guide data reduction procedure.

5. Numerical studies for influence on flow and heat transfer

One of the leading numerical investigations about the OSF heat exchangers is carried out by Joshi and Webb [3]. The laminar and turbulent flow regimes are analytically modeled to predict the *j* and *f* factors of OSF. Moreover, the experiments are employed to visualize the flow structure, transition regime in particular. The ultimate purpose of this endeavor is to classify the regimes as the laminar, transition, or turbulent as best as it could be. The necessary equations for *Nu* and *f* are derived from energy and momentum balances. The character of

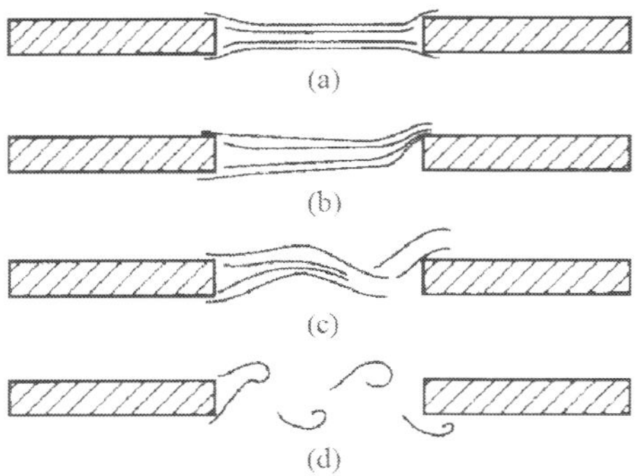

Figure 9. Flow patterns observed in visualization experiments [3].

flows on fins and in wakes is illustrated (**Figure 9**) with regard to the data attained from the numerical analysis, which are demonstrated with respect to velocity profiles. Since the effect of the fin length, thickness and spacing on the wake have not revealed in earlier communications, it is focused to find out this by either visualization or numerical analysis. The findings of the experiments are demonstrated by **Figure 9**, which starts to oscillate when the regime turns from laminar to turbulent (from (a) to (d) in **Figure 9**). Parametric study is included as well and the results of the friction factor are summarized. It is noted that when the thickness to length ratio ascends, the friction factor increases [3]. A correlation (**Table 3**) that bases on the wake width is developed to predict the *Re* at the transition points from the *Re* numbers of 21 different heat exchangers reported in literature.

Along the same path, the effect of thickness for the flow regime and performance is examined numerically [17]. And the flow pattern is investigated as presented in **Figure 10** (fin thickness ascends from top to bottom cartoon). With regard to that results when the fin is thin enough no particular difference could be observed from the low *Re* number. With the increment of the fin thickness recirculation zones emerge next to the plate.

Another study about the effect of parameters on the performance of the offset strip fin was reported by Kim et al. [2]. Even though the ultimate purpose of the study is to provide a better correlation for the offset strip fins at higher blockage ratios and different fluids, the parametric effect is also observed. It is remarked that the j and f factors descend when the spacing and length increases and in contrary it rises when the thickness increases. By the means of pressure drop, a similar trend could be observed with the j and f factors. It is highlighted that the thickness effect is more prominent on the performance than the spacing and length.

Even though most of the studies employed for the flow at horizontal heat exchangers, Suzuki et al. [18] operated the flow at vertical flat tubes used in the free-forced-mixed convection at low *Re*. The investigation has proceeded for a staggered array using numerical and experimental approaches. Three rows of flat plates are arranged in a staggered pattern where the air flows upward at a low speed.

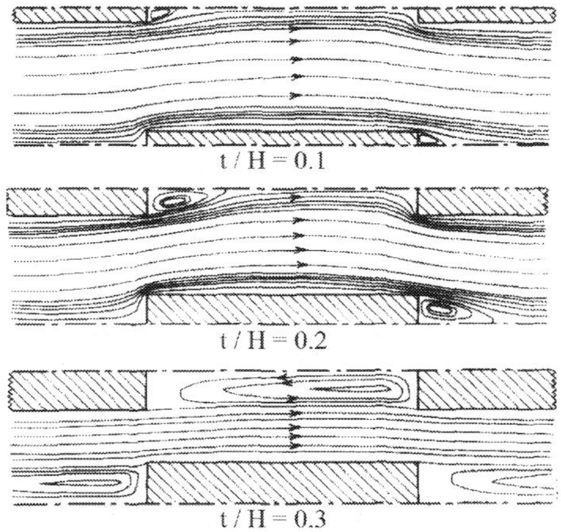

Figure 10. Flow patterns for different plate thickness [17].

In the numerical computations, two methods are used: the first one is α-ω method, the second one is U-V-P method (which solves axial and normal velocity components U and V together with pressure correction ΔP); in order to see the difference, both methods are run for the same case where it is seen they are almost identical. The effect of thickness is observed along the flow

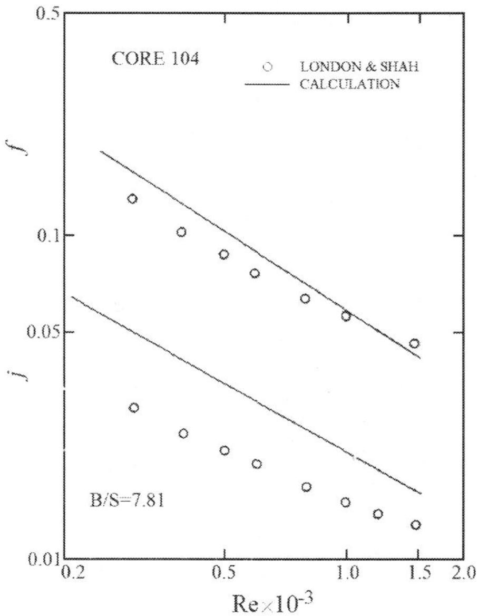

Figure 11. Comparison of the experiments by London and Shah [18].

length in the described array and any significant impact could be observed at the slower flow regimes. In all those trials, the highest heat transfer is obtained at the entrance of the array. A comparison between the published data from London and Shah [6] and the numerical data of the study is presented (**Figure 11**). It is noted that there is a good agreement between the calculated results and London and Shah for their offset geometry Core 104. Finally, it is concluded that the thicker fins are not necessary for the slower regimes since there is no distinctive effect on the performance. Even though the shorter fins give better heat transfer performance due to its higher production cost it is not efficient to replace with longer fins.

The physical phenomena for the heat transfer mechanism in offset strip fin geometries for the self-sustained oscillatory flow at high *Re* range is also considered by the researchers. Saidi et al. [19] noted that, due to the nature of the compact heat exchangers, in most of the studies, the flow regime is at lower *Re* range; in addition, in some rare applications, some researchers examine the flow at higher *Re* range in order to see the response at the turbulent. The flow regime at the intermediate *Re* range is not very common and there are unrevealed parts for this flow type. A numerical study based on finite volume method computed by a CFD code is given in the communication [20]. Friction factor and Colburn *j*-factor are compared by DeJong

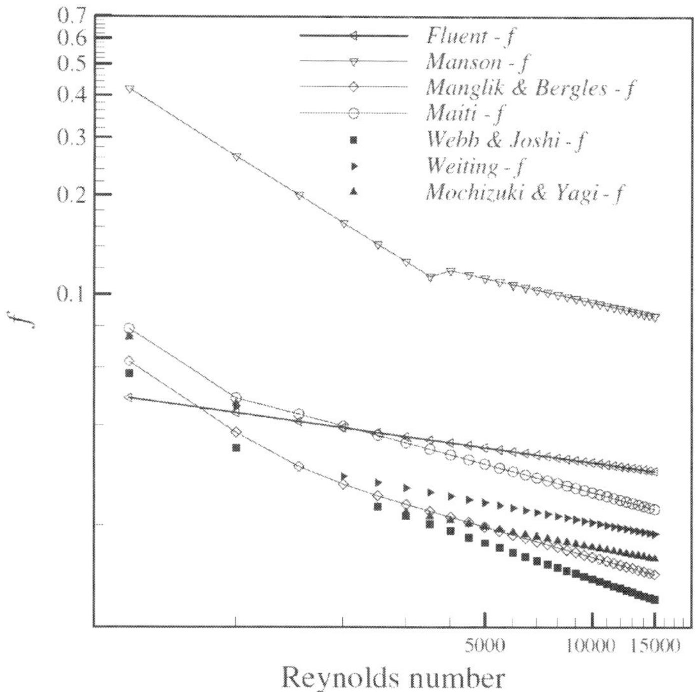

Figure 12. The comparison of Colburn *j*-factor of different heat transfer surface types [21].

et al. [20] for the confirmation of the findings. Then, the velocity field around the fin during a complete period of oscillation is presented by contour plots for various time steps in the sequence of development. Moreover, unsteadiness is presented in the same manner. The second moment correlation of fluctuating velocity components and second moment of temperature velocity fluctuations are demonstrated. It is noted that heat transfer and moment transfer are dissimilar.

Orientation of inlet and outlet headers plays a major role in performance of compact heat exchangers, especially in aerospace applications where the orientations are not straight. In order to understand the maldistribution at these components, Ismail et al. [21] studied these phenomena numerically. Three headers are developed to modify the distribution, which are combined with three offset strip fins and sixteen wavy fin geometries used in the heat exchangers. The computational results for wavy and offset strip fin structures are validated by analytical results and a good agreement is observed for *j* within –2% and for *f* within –9%, as it is illustrated in **Figures 12** and **13**. As for the headers, the flows are analyzed for either real (without modified headers) or ideal cases (with modified headers) and in all these conditions, the pressure drop is higher at real cases than at the ideal case. Among those, the biggest real case-ideal case difference is observed for the heat exchanger with baffle plate by 34%.

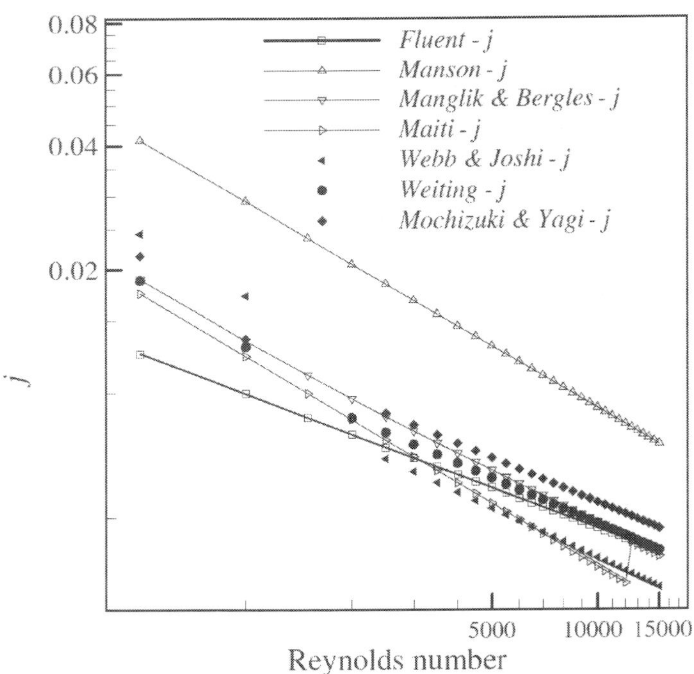

Figure 13. The comparison of friction factor of different heat transfer surface types [21].

6. Alternate applications benefit from offset strip fins

Even though the following applications are out of scope of this study, it is worth to note these in order to introduce the alternate ones to the readers who are not aware of. Solar air heater is one of these alternate systems that can transfer solar irradiant energy into airflow heat. These systems are typically used in space heating and food processing and they become popular for their cost-effectiveness and easy maintenance [22]. Like all other thermal systems, the development and research on these systems continue. In order to improve the performance of solar air heaters, researchers considered the use of offset strip fins [22–25] either numerically or experimentally. Results showed that, instantaneous thermal efficiency exceed 0.40 even at the lower air flow rates [22]. High thermal performance is obtained with low friction and as a result of all those power consumption of the fan descends [23]. In the laminar flow regime, the greatest temperature variation is recorded as 7°C at Re =479 [24]. On another study, the total amount of the collected heat is 612 W/m^2 for an incident heat flux of 900 W/m^2 [25].

One of the other applications that uses offset strip fin is oil cooled heat exchangers where the need to improve heat transfer at the oil side is vital for the performance of the entire system. Due to the nature of the heat transfer medium, the fins are made up of steel instead of aluminium. The correlations for the system are derived from the experiments and compared with the experimental data; with regard to that, heat transfer is ±10% within the experiments with a mean deviation of 4.01%, while the friction factor can describe the experimental data within ±15% with a mean deviation 5.68% at low Re number [26]. Thermal hydraulic characteristics of offset strip fin are examined by considering the flow angle. Varying angles between 0 and 90° are observed and among those, the highest performance is observed at the angle 45° [27].

Ice slurry applications with its high transportation capacity by the latent heat, emerges as a promising alternate for the single phase coolant [28]. Pressure drop and heat transfer coefficient are investigated as common as in the OSF studies. But unlike the common approach, partial thermal resistance is examined as well. An empirical correlation for the Colburn j-factor is derived as well. The results revealing the instabilities of the flow may be seen due to total or partial blockages at isothermal conditions. In addition, a slight increase in the heat transfer rate and in the overall heat transfer coefficient is noted when the ice fraction increases at the flow.

Another two phase flow case is studied by some other researchers [29] for OSF heat exchangers. A vertical adiabatic channel OSF is tested for R113 (hydrofluorocarbon (trichloro-1,1,2-trifluoro-1,2,2-ethane), which is a refrigerant and now phased out in most of the developed countries) and not only pressure drop is measured but also two phase flow is visually photographed. A significant enhancement (40–50% higher) is observed for the two phase flow of R113 in OSF matrix than the conventional round tube.

7. Correlations of j and f factors in offset strip fin heat exchangers

The reliable prediction of the heat transfer and pressure drop in offset strip fin is vital and very hard to achieve. Various correlations have been developed by numerous researchers in decades to succeed and fulfill this requirement. Since most of the offset strip fin studies focus to reveal the effect of the geometrical parameters, the large amount of data produced in the researches belong to parametric investigations. Most of the *j* and *f* correlations that could be found in literature bases on geometrical parameters as given in **Table 3**. The nondimensional parameters used in the correlations are, $\alpha=s/h$, $\delta= t/l$ and $\gamma=t/s$ where length is *l*, height is *h*, transverse spacing *s* and thickness *t* as noted before.

A particular comparison with respect to the Kays and London's experimental study [5] is taken to the following to see the differences between the approaches for the same particular problem. In the given comparison, the correlations pertain to Wieting [30], Joshi and Webb [4] and Muzychka et al. [31] are examined for a particular fin structure (uninterrupted fin length is

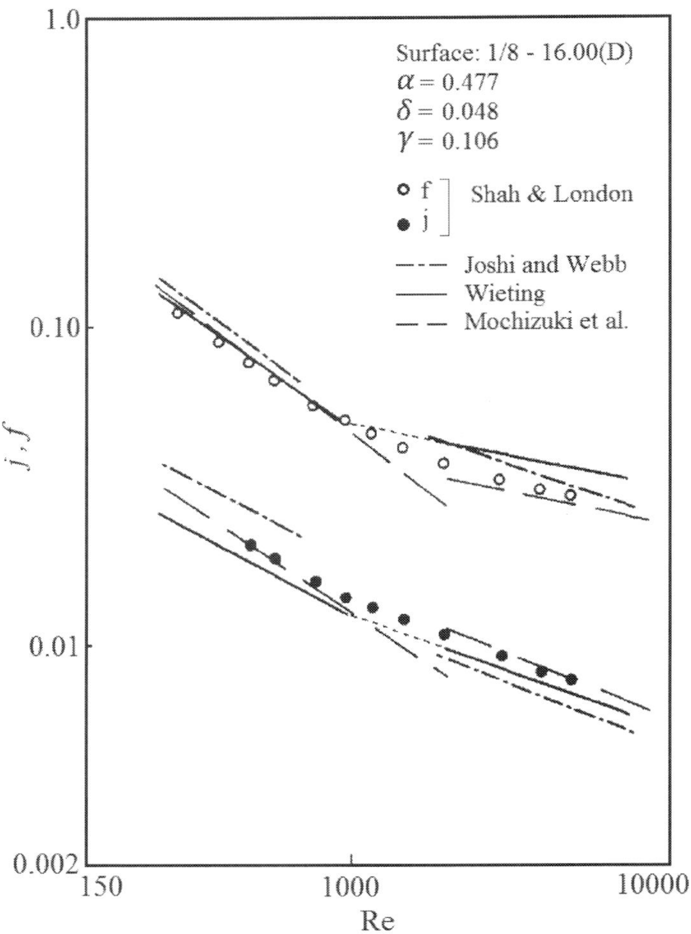

Figure 14. Comparison of *j* and *f* correlations with experimental data of Kays and London [4].

1/8 inch—number of fins per inch is 16.00 and double sandwiched fin structure) from Kays London experiments [3]. In **Figure 14**, it could be seen that, all three correlations, the most cited correlations in literature for this fin type can predict the *j* and *f* values in a good agreement, but they are not sufficient to present the transition between laminar and turbulent.

That inadequacy leads the researchers to find out a correlation, which can meet and correspond to all regimes at once. Manglik and Bergles tried to meet this with the correlation given in **Table 3**, which is claimed as giving a continuous solution for the flow regimes turning from laminar to turbulent. The calculations for the same, Kays and London problem (fin structure (1/8—16.00 D)), is also reported in the study (**Figure 15**).

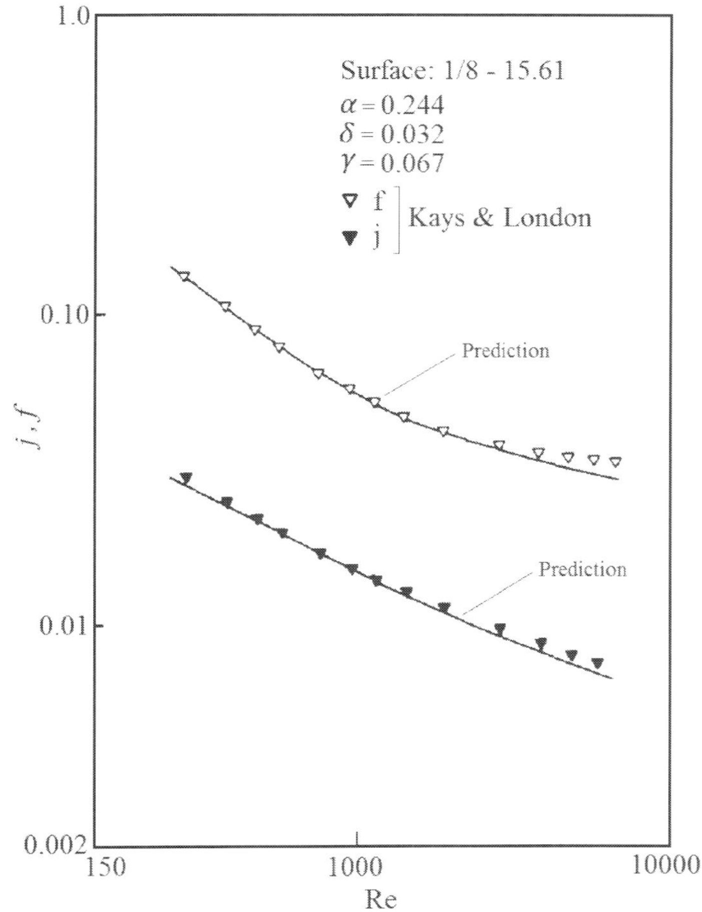

Figure 15. Comparison of correlation with experimental data of Kays and London [4].

Most of the correlations that could be met in literature are derived from the experimental studies, but unlike the common approach, some researchers try to extract the correlations from their numerical experience such as Kim et al. [2]. In the study, 39 different models are analyzed numerically by commercial CFD software where different geometric parameters and varying fluids are tried. It is also underlined that the earlier correlations are not sufficient to estimate the Colburn *j*-factor and friction factor when the blockage ratio (β) increases (β =12–27%). So there is a demand to find the answer of that question. The analysis are carried out by different turbulent models and the most suitable one is decided by comparing the findings with the most preferred correlations in literature [3, 4] and then the remaining calculations are performed to obtain a new correlation to attain the best fitting one not only for different fluids but also for different blockage ratios. The correlations are grouped as regarding to their blockage ratio (β). Since the blockage ratios below 20% are not efficient for offset strip fins and very high pressure drop could be seen beyond 35%, the blockage range investigated in the study is taken as 20–35%, the Colburn *j*-factor and friction *f*-factor can be correlated as in **Table 3**.

As mentioned earlier, the blockage ratio is not the only objective of the study [2], but the Prandtl effect is also observed by the usage of different fluids such as water; the correlations according to Prandtl can also be found as in the lower part of **Table 3**. These correlations are compared by the Manglik and Bergles for the blockage ratio of 12 and 27% in

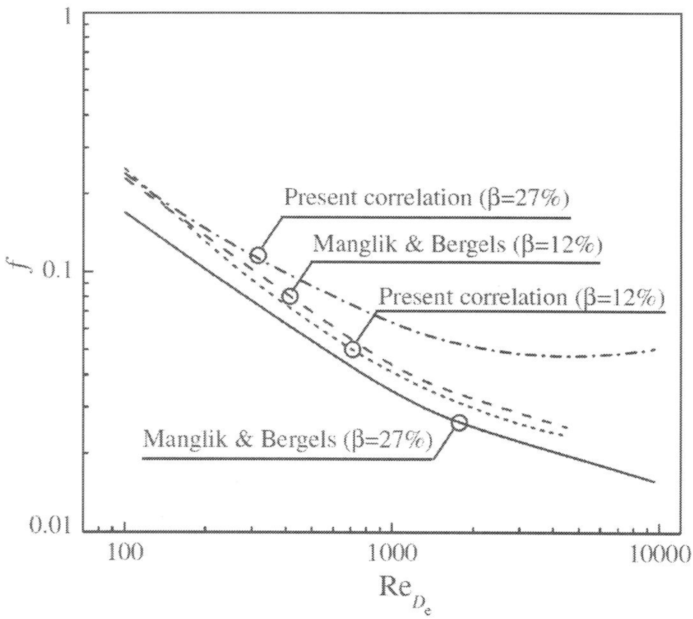

Figure 16. Comparison of Kim et al. correlation and Manglik and Bergles correlation for the blockage ratio of β = 12% and β = 27% for friction factor [2].

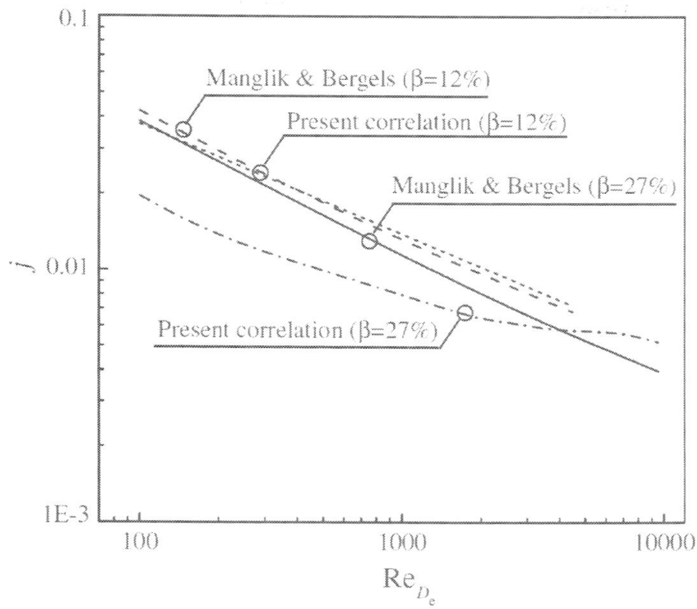

Figure 17. Comparison of Kim et al. correlation and Manglik and Bergles correlation for the blockage ratio of β = 12% and β = 27% for Colburn factor [2].

Figures 16 and **17**. As it could be seen, the M&B correlation starts to overestimate when the blockage ratio is over 20%.

8. Concluding remarks

In this particular chapter, the ultimate purpose is to inform and address in detail about the studies on offset strip fins heat exchangers that have been performed by the researchers for decades and also to call a strong attention to the superior advantages of their use. The basics of this structure are reported; in addition, the investigations found in the literature are summarized by considering their objectives. The parametric effects of the structural, experimental and numerical research of the fin under varying flow regimes and conditions, the evolution of heat transfer and friction factors under different flow conditions are all given with regard to the objective of the investigations, which are cited. The outcomes that have to be highlighted at the end of this section can be listed as in the following:

- The highest heat transfer is observed at the entrance of the gate where the highest dislocations emerged.
- The shorter fin length leads to more interrupted flow, which provides a better heat transfer of this particular fin geometry.
- The Colburn factor *j* descends and the friction factor *f* ascends by the increase of the fin pitch.

- Both of these factors (j and f) increase with the decrease of fin length.
- Colburn factor j-factor decreases, while the friction factor f increases when fin bending distance decreases.
- The flow streams could be illustrated depending on the calculated data, which provide to visualize the regime evolution. In addition, the recirculation zones emerge with the increment of the thickness.

Other performance criteria derived and used for the offset strip fins are flow area goodness factor j/f, the ratio $j/f^{1/3}$ and thermal hydraulic performance factor JF. With regard to the investigations:

- JF factor could be useful for the water while $j/f^{1/3}$ is more convenient for gas-oil liquid. In contrary to the given two factors, j/f couldn't be considered as a good performance criterion for fluids.

The reliable prediction of the heat transfer and pressure drop in offset strip fin is crucial. The correlations derived in the studies are presented together in **Table 3** in order to provide a better comparison.

- With regard to the coverage, good agreement with the experimental data and mostly cited correlation pertains to Manglik and Bergles among the given equations.

Nomenclature

A_c	Heat transfer area of the cold side, m^2
$A_{c,f}$	Cross-sectional area of the fin, m^2
A_f	Frontal area, m^2
A_h	Heat transfer area of the hot side, m^2
A_t	Heat transfer area of the tube, m^2
C_f	Fin bending distance, mm
$c_{p,c}$	Specific heat of the cold stream, kJ/(kg °C)
$c_{p,h}$	Specific heat of the hot stream, kJ/(kg °C)
C_r	Heat capacity ratio
D	Diameter, mm
D_e	Hydraulic diameter, mm
f	Fanning friction factor
f_R	Reference value of the friction factor
h	Height of the fin, mm
h_c	Heat transfer coefficient of the cold (external) stream, W/(m^2 °C)
h_h	Heat transfer coefficient of the hot (internal) stream, W/(m^2 °C)

j	Colburn j-factor
j_R	Reference value of the Colburn j-factor
JF	Performance evaluation criteria related with volume goodness factor
k	Thermal conductivity of the material, W/(m °C)
k_c	Pressure loss coefficient at the inlet of the heat exchanger
k_e	Pressure loss coefficient at the exit of the heat exchanger
k_t	Thermal conductivity of the tube material, W/(m °C)
l	Uninterrupted fin length, mm
m	Parameter for the calculation of η_f related with h_c, $A_{c,f}$, P_f and k, m^{-1}
$\dot{m}_c$	Mass flow rate of the cold stream, kg/s
$\dot{m}_h$	Mass flow rate of the hot stream, kg/s
N_s	Entropy generation number
Nu	Nusselt number
NTU	Number of transfer unit
$NSTD$	Normalized standard deviation
P	Pressure, Pa
P_f	Perimeter of the fin, m
Pr	Prandtl number
$\dot{Q}$	Average heat transfer rate, W
$\dot{Q}_c$	Heat transfer rate of the cold stream, W
$\dot{Q}_h$	Heat transfer rate of the hot stream, W
$\dot{Q}_{max}$	Maximum heat transfer rate, W
Re	Reynolds number
Re_{D_e}	Reynolds number based on the equivalent diameter
s	Transverse spacing, mm
t	Thickness of the fin material, mm
T	Temperature, °C
$T_{c,i}$	Inlet temperature of the cold stream, °C
$T_{c,o}$	Outlet temperature of the cold stream, °C
$T_{h,i}$	Inlet temperature of the hot stream, °C
$T_{h,o}$	Outlet temperature of the hot stream, °C
ΔT_m	Logarithmic mean temperature difference, °C
St	Stanton number
STD	Standard deviation
u	Free velocity of the external fluid, m/s
U	Overall heat transfer coefficient, W/m^2 °C
UA	Overall thermal conductance, W/°C

Greek letters

α	Nondimensional parameter, *s/h*
β	Blockage ratio
γ	Nondimensional parameter, *t/s*
δ	Nondimensional parameter, *t/l*
δ_t	Thickness of the flat tube material, mm
ε	Effectiveness
η_c	Surface effectiveness of the cold side
η_f	Fin efficiency
μ	Dynamic viscosity, kg/ms
υ	Kinematic viscosity, m^2/s
ρ	Density, kg/m^3
λ	Thermal conductivity of the fluid, W/(m °C)
ψ^*	Performance evaluation criteria

Author details

Latife Berrin Erbay[1]*, Mehmet Mete Öztürk[2] and Bahadır Doğan[1]

*Address all correspondence to: lberbay@ogu.edu.tr

1 Department of Mechanical Engineering, Faculty of Engineering and Architecture, Eskisehir Osmangazi University, Eskisehir, Turkey

2 Department of Motor Vehicles and Transportation Technology, Vocational School of Transportation, Anadolu University, Eskisehir, Turkey

References

[1] Kakaç S, Liu H. Heat Exchangers: Selection, Rating and Thermal Design. 2nd ed. Boca Raton: CRC Press; 2002.

[2] Kim M-S, Lee J, Yook S-J, Lee K-S. Correlations and optimization of a heat exchanger with offset-strip fins. International Journal of Heat and Mass Transfer. 2011;**54**(9–10):2073–2079. DOI: 10.1016/j.ijheatmasstransfer.2010.11.056

[3] Joshi H M, Webb R L. Heat transfer and friction in the offset strip-fin heat exchanger. International Journal of Heat and Mass Transfer. 1987;**30**(1):69–84. DOI: 10.1016/0017–**9310**(87)90061–5

[4] Manglik R M, Bergles A E. Heat transfer and pressure drop correlations for the rectangular offset strip fin compact heat exchanger. Experimental Thermal and Fluid Science. 1995;**10**(2):171–180. DOI: 10.1016/0894–**1777**(94)00096-Q

[5] Kays W M, London A L. Compact Heat Exchangers. 3rd ed. New York: McGraw-Hill; 1984.

[6] London A L, Shah R K. Offset rectangular plate-fin surfaces - Heat transfer and flow friction characteristics. Trans. ASME, Journal of Engineering for Power. 1968;**90**(3):218–228. DOI: 10.1115/1.3609175

[7] Bhowmik H, Lee K-S. Analysis of heat transfer and pressure drop characteristics in an offset strip fin heat exchanger. International Communications in Heat and Mass Transfer. 2009;**36**(3):259–263. DOI: 10.1016/j.icheatmasstransfer.2008.11.001

[8] Khoshvaght-Aliabadia M, Hormozia F, Zamzamian A. Role of channel shape on performance of plate-fin heat exchangers: Experimental assessment. International Journal of Thermal Sciences. 2014;79:183–193. DOI: 10.1016/j.ijthermalsci.2014.01.004

[9] Yujie Y, Yanzhong L, Biao S, Jieyu Z. Performance evaluation of heat transfer enhancement in plate-fin heat exchangers with offset strip fins. Physics Procedia. 2015;67:543–550. DOI: 10.1016/j.phpro.2015.06.073

[10] Dong J, Chen J, Chen Z, Zhou Y. Air-side thermal hydraulic performance of offset strip fin aluminum heat exchangers. Applied Thermal Engineering. 2007;**27**(2–3):306–313. DOI: 10.1016/j.applthermaleng.2006.08.005

[11] Peng H, Ling X, Juan L. Performance investigation of an innovative offset strip fin arrays in compact heat exchangers. Energy Conversion and Management. 2014;80:287–297. DOI: 10.1016/j.enconman.2014.01.050

[12] Fernández-Seara J, Diz R, Uhía F J. Pressure drop and heat transfer characteristics of a titanium brazed plate-fin heat exchanger with offset strip fins. Applied Thermal Engineering. 2013;**51**(1–2):502–511. DOI: 10.1016/j.applthermaleng.2012.08.066

[13] Michna G J, Jacobi A M, Burton R L. Friction factor and heat transfer performance of an offset-strip fin array at air-side Reynolds numbers to 100,000. In: International Refrigeration and Air Conditioning Conference; School of Mechanical Engineering, Purdue University. West Lafayette: Purdue e-Pubs; 2006.

[14] Saad S B, Clement P, Gentric C, Fourmigue J-F, Leclerc J-P. Experimental distribution of phases and pressure drop in a two-phase offset strip fin type compact heat exchanger. International Journal of Multiphase Flow. 2011;**37**(6):576–584. DOI: 10.1016/j.ijmultiphase flow.2011.03.009

[15] Hu S, Herold K E. Prandtl number effect on offset fin heat exchanger performance: experimental results. International Journal of Heat and Mass Transfer. 1995;**38**(6):1053–1061. DOI: 10.1016/0017–**9310**(94)00220-P

[16] Hu S, Herold K E. Prandtl number effect on offset fin heat exchanger performance: predictive model for heat transfer and pressure drop. International Journal of Heat and Mass Transfer. 1995;**38**(6):1043–1051. DOI: 10.1016/0017–**9310**(94)00219-L

[17] Patankar S V, Prakash C. An analysis of the effect of plate thickness on laminar flow and heat transfer in interrupted-plate passages. International Journal of Heat and Mass Transfer. 1971;**24**(11):1801–1810. DOI: 10.1016/0017–**9310**(81)90146–0

[18] Suzuki K, Hirai E, Miyake T. Numerical and experimental studies on a two-dimensional model of an offset-strip-fin type compact heat exchanger used at low Reynolds number. International Journal of Heat and Mass Transfer. 1985;**28**(4):823–836. DOI: 10.1016/0017–**9310**(85)90232–7

[19] Saidi A, Sunden B. A numerical investigation of heat transfer enhancement in offset strip fin heat exchangers in self? sustained oscillatory flows. International Journal of Numerical Methods for Heat & Fluid Flow. 2001;**11**(7):699–716. DOI: 10.1108/EUM0000000005984

[20] DeJong N C, Gentry M C, Jacobi A M. An entropy-based, air-side heat exchanger performance evaluation method: Application to a condenser. HVAC...R Research. 1997;**3**(3):185–195. DOI: 10.1080/10789669.1997.10391373

[21] Ismail L S, Ranganayakulu C, Shah R K. Numerical study of flow patterns of compact plate-fin heat exchangers and generation of design data for offset and wavy fins. International Journal of Heat and Mass Transfer. 2009;**52**(17–18):3972–3983. DOI: 10.1016/j.ijheatmasstransfer.2009.03.026

[22] Yang M, Yang X, Li X, Wang Z, Wang P. Design and optimization of a solar air heater with offset strip fin absorber plate. Applied Energy. 2014;113:1349–1362. DOI: 10.1016/j.apenergy.2013.08.091

[23] Hachemi A. Experimental study of thermal performance of offset rectangular plate fin absorber-plates. Renewable Energy. 1999;**17**(3):371–384. DOI: 10.1016/S0960–**1481**(98)00115–3

[24] Youcef-Ali S, Desmons J Y. Numerical and experimental study of a solar equipped with offset rectangular plate fin absorber plate. Renewable Energy. 2006;**31**(13):2063–2075. DOI: 10.1016/j.renene.2005.10.008

[25] Youcef-Ali S. Study and optimization of the thermal performances of the offset rectangular plate fin absorber plates, with various glazing. Renewable Energy. 2005;**30**(2):271–280. DOI: 10.1016/j.renene.2004.04.009

[26] Guo L, Chen J, Qin F, Chen Z, Zhang W. Empirical correlations for lubricant side heat transfer and friction characteristics of the HPD type steel offset strip fins. International Communications in Heat and Mass Transfer. 2008;**35**(3):251–262. DOI: 10.1016/j.icheatmasstransfer.2007.07.006

[27] Guo L, Qin F, Chen J, Chen Z. Lubricant side thermal-hydraulic characteristics of steel offset strip fins with different flow angles. Applied Thermal Engineering. 2008;**28**(8–9):907–914. DOI: 10.1016/j.applthermaleng.2007.07.005

[28] Fernández-Seara J, Diz R. Thermo-hydraulic behavior of ice slurry in an offset strip-fin plate heat exchanger. International Journal of Refrigeration. 2014;41:171–180. DOI: 10.1016/j.ijrefrig.2013.12.011

[29] Mandrusiak G D, Carey V P. A Method for detecting liquid shedding during annular flow in channels with offset strip fins. Experimental Thermal and Fluid Science. 1991;**4**(2):239–245. DOI: 10.1016/0894-**1777**(91)90068-3

[30] Wieting A R. Empirical correlations for heat transfer and flow friction characteristics of rectangular offset-fin plate-fin heat exchangers. Journal of Heat Transfer. 1975;**97**(3):488–490. DOI: 10.1115/1.3450412

[31] Muzychka Y S, Yovanovich M M. Modeling the f and j characteristics of the offset strip fin array. Journal of Enhanced Heat Transfer. 2001;**8**(4):261–277. DOI: 10.1615/JEnhHeatTransf.v8.i4.40

Chapter 4

Comprehensive Study of Heat Exchangers with Louvered Fins

Latife Berrin Erbay, Bahadır Doğan and
Mehmet Mete Öztürk

Additional information is available at the end of the chapter

http://dx.doi.org/10.5772/66472

Abstract

The purpose of this chapter is to describe the basic physical features and the analysis of the thermal-hydraulic performance of the louver fins. The terminology which is used widely in the field of compact heat exchanger with louvered fin is described. The flow phenomenon affected by the operating conditions and the geometric parameters of the louvered fin is examined using the flow visualization techniques found in the literature. A methodology is given to calculate the heat transfer and the friction factor. Stanton number, Colburn *j*-factor and friction factor are defined as a performance criteria and the variations of these criteria with respect to the Reynolds number and the geometric parameters of the louvered fin. The combinations of these dimensionless number such as area goodness factor (j/f), volume goodness factor ($j/f^{1/3}$) and *JF* number related with the volume goodness factor are discussed in terms of overall performance criteria. Finally, the correlations of the louvered fin heat exchanger and their tabulated data was summarized.

Keywords: plate fin, louver fin, compact heat exchanger

1. Introduction

The plate heat fin exchangers have a widely used application area. They are characterized by having secondary surfaces or fin structures. The function of the secondary (extended) surface is the enhancement of heat transfer performance of the heat exchanger in the allowable range of the pressure drop. The commonly used forms of the extended surface of the plate fin heat exchangers are the triangular or rectangular plain fin, offset strip fin, wavy fin, louvered fin and perforated fin as shown in **Figure 1**.

For extended surface application, fin geometries fall into two categories: continuous and interrupted surfaces. Continuous surfaces achieve heat transfer enhancement through the

INTECH
open science | open minds

© 2017 The Author(s). Licensee InTech. This chapter is distributed under the terms of the Creative Commons Attribution License (http://creativecommons.org/licenses/by/3.0), which permits unrestricted use, distribution, and reproduction in any medium, provided the original work is properly cited.

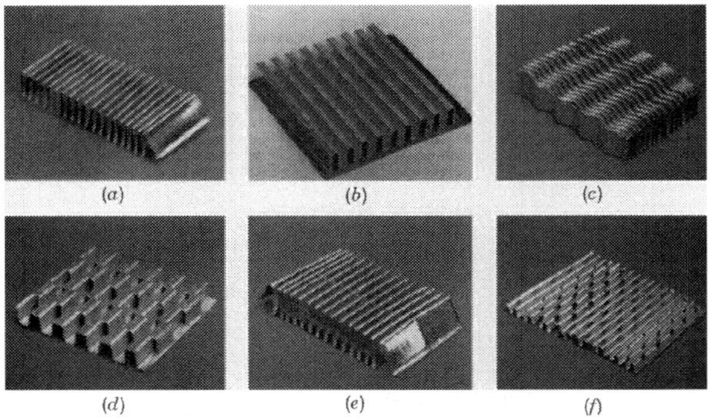

Figure 1. Fin geometries for plate fin heat exchangers: (a) plain triangular fin; (b) plain rectangular fin; (c) wavy fin; (d) offset strip fin; (e) multi-louver fin; (f) perforated fin [1].

secondary flow patterns introduced by sudden velocity changes. On the other hand, interrupted surfaces achieve heat transfer enhancement by the continuous growth and destruction of laminar boundary layers on the interrupted portion of the geometry. One of the mostly used example of the interrupted surface is the louvered fin. The louvered fins were firstly investigated by Kays and London [2] in 1950s, and the popularities of the louvered fins have been maintained.

Today, the use of louvered fins has become popular in the fields of automotive, heating, cooling, air conditioning, power plants and food industry. Typical structure of the louvered fin is shown in **Figure 2**. The efforts of maximize the heat transfer and minimize the pressure drop in heat exchanger design are rapidly increasing due

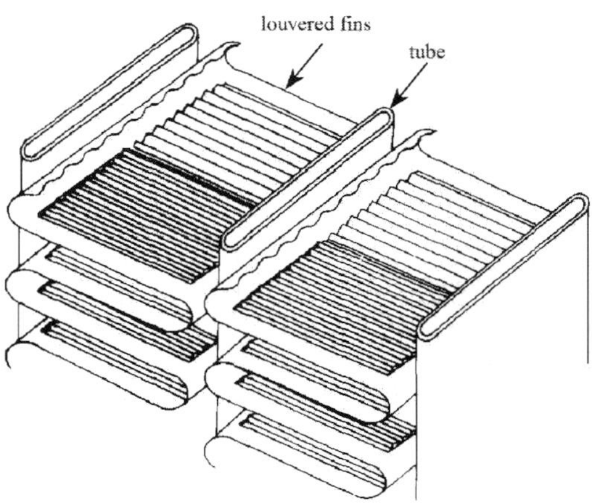

Figure 2. Typical structure of a louver fin geometry [4].

to the restrictions of energy consumption applied by the governments. In this case, the importance of light, high surface density and energy efficient heat exchangers is increasing. The louvered fins are commonly used in heat exchanger field for reasons beyond simply increase the heat transfer surface area and decrease the volume, the amount of coolant and the costs [3].

The louvered fins enhance the heat transfer by providing multiple flat-plate leading edges with their associated high values of heat transfer coefficient. Although the louvered fins are similar in principle to the offset strip fin, they can enhance heat transfer by a factor of 2 or 3 compared with equivalent un-louvered surfaces. The louvers have the further advantage that the enhancement of heat transfer is gained without increase in flow resistance that results from the use of turbulators [5].

Louvers are generally formed by cutting the metal and pushing out the cut elements from the plane of the base metal. They can be manufactured by high-speed production techniques and as a result are less expensive than other interrupted flow geometries when produced in large quantities. Louvered fin geometries can be made in different dimensions of fin length, louver length or material thickness depending on fabrication techniques [2]. The physical principle of the louvered fin is based on the breaking the boundary layer of the flow that passes through the louvers. The structure of the flow phenomenon over the louvered fins is given in details at the following section.

Due to the extensive use of the louvered fins in the heat exchanger area, the researches have spent great efforts to improve the louvered fin geometry from 1950s. In this chapter, first the terminology and the fundamental concepts of flow phenomenon of the louvered fin are explained. In the following sections, the heat transfer and pressure drop characteristics of the louver fins are examined with respect to the geometrical variations of the louvered fin geometry adhering to the experimental and numerical studies in the literature. The empirical correlations are also summarized by a tabulated data.

2. Terminology of the louvered fin

The terminology of a louvered fins was firstly created by Kays and London [2] as shown in **Figure 3**. Each louvered fin configurations is designated by two figures. The first indicates the length of the louvered fin in the flow direction and the second indicates the fin pitch per inch transverse to the flow. Therefore, the meaning of "1/2–6.06" is that each louver has length of 1/2 inch in the flow direction and 6.06 fins in per inch.

Another terminology used for louvered fins with a flat tube is illustrated in **Figure 4**. It is shown from the cross-sectional view that the gap between two louvered fins is called fin pitch (F_p). The length of the flow from the leading edge up to the end of the fin is called flow depth (F_d). The vertical length of the fin and the louver are called fin height and louver height and designated by F_h and L_h, respectively. The horizontal length of the gap between each louver is called louver pitch (L_p), and each louver has an angle of L_α.

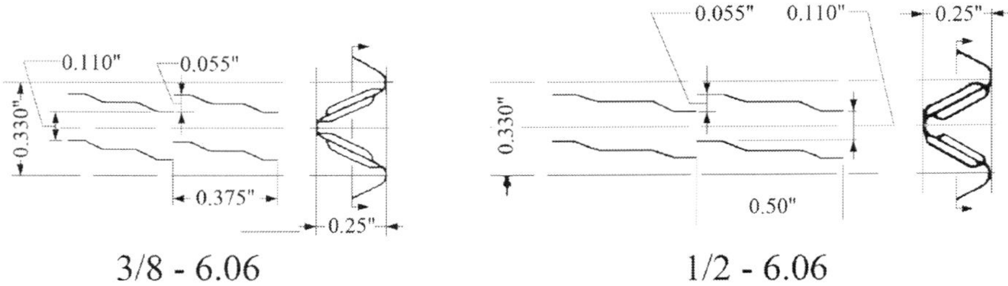

Figure 3. Terminology of the louvered fin [2].

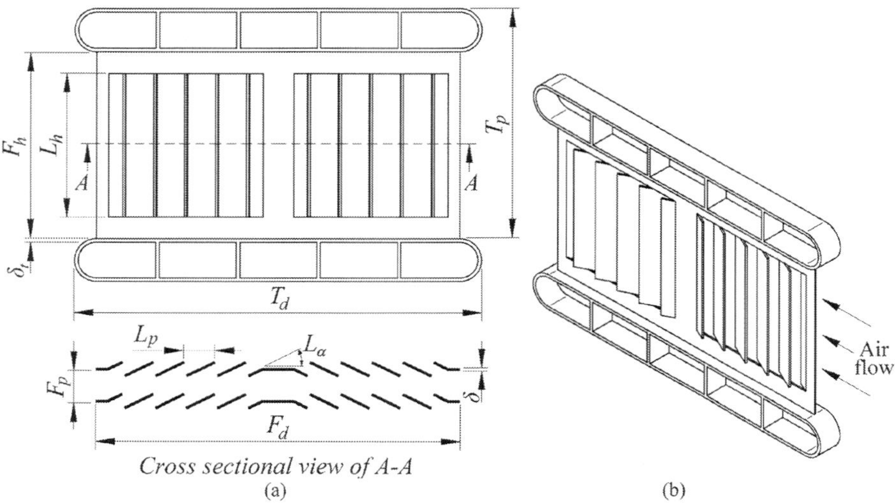

Figure 4. Terminology of the louvered fin [6].

3. Flow phenomenon in louvered fin arrays

The structure of the flow passes through the louvered fins can be identified with flow visualization technique using dye injection and hydrogen bubble. Numerical analysis is also a method to identify the flow phenomenon of louvered fins. Several experimental and numerical studies indicate that the geometric design and the free velocity of the flow effect the direction of the flow through the louvers. As shown in **Figure 5**, the flow enters from the leading edge of the first louver and then directed by the louvers or the fins. If the greatest proportion of the flow is passing between the louvers, it is called "louver directed flow". Similarly, if the greatest proportion of the flow is passing through the gap between the fins, it is called "fin or duct directed flow" [7].

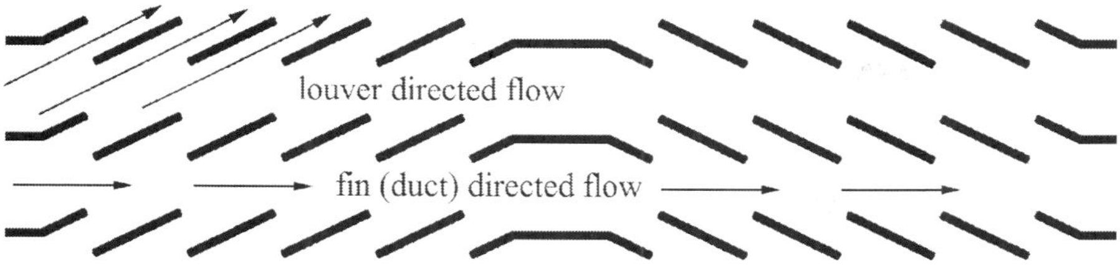

Figure 5. Possible flow direction of flow through the louvered fins.

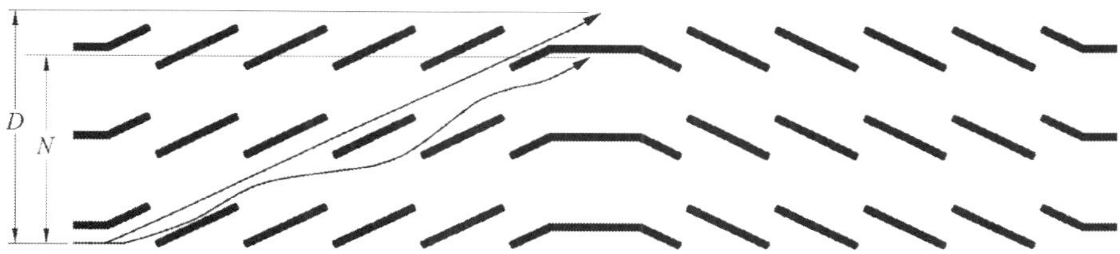

Figure 6. Definition of the flow efficiency.

This definition has revealed another concept called flow efficiency (η) [3]. **Figure 6** provides a visual definition of the flow efficiency, and it is calculated by Eq. (1)

$$\eta = \frac{N}{D} = \frac{Actual\ transverse\ distance}{Ideal\ transverse\ distance} \tag{1}$$

According to the Eq. (1), if the flow efficiency is equal to 1, the flow is parallel to the louvers. If it is equal to zero, the flow is axial through the louvered fin array which is 100% duct flow [3]. **Figure 7** shows the structure of the flow between the louvers. The photographs of the flow through the louvered fin for a louver angle of 26° at a Reynolds number of 500 are obtained using dye injection and hydrogen bubbles techniques. It can be seen that the significant proportion of the flow is directed by the louvers under these circumstances. It is observed that the boundary layers exist on both the upper and lower surfaces of the louver. Flow separation is observed on the back side of the inlet louvers and boundary layer exists on both the upper and lower surfaces of the louvers [3].

Additionally, a large adverse pressure gradient exists on the downstream side of the louvered fin, and the flow separates at the leading edge and forms a large recirculation bubble. **Figure 8a** clearly shows this recirculation zone which is called "first recirculation zone" for a louver angle of 25° at a Reynolds number of 510 using naphthalene sublimation technique by DeJong and Jacobi [8]. This large recirculation zone circulates in a clockwise direction, resulting in a region of very high shear near the trailing edge of the louver where flow that passes downstream between fins interacts with the separation bubble. The high shear results in the formation of a small "secondary recirculation zone" with counter-clockwise rotation in the wake just downstream of the end of the louver. Except at very low Reynolds numbers, a

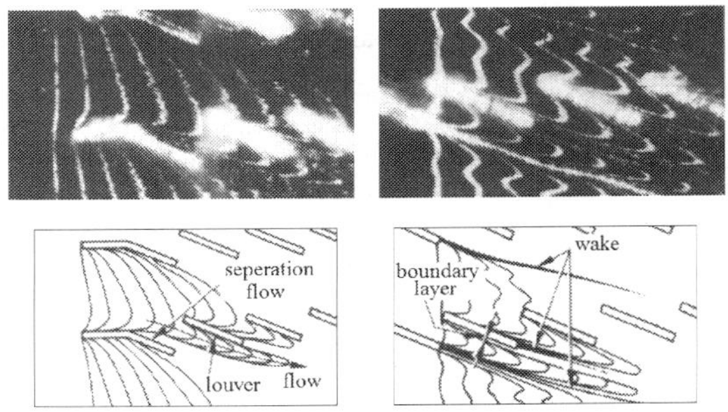

Figure 7. Visualization of the flow in the louvered fin array [3].

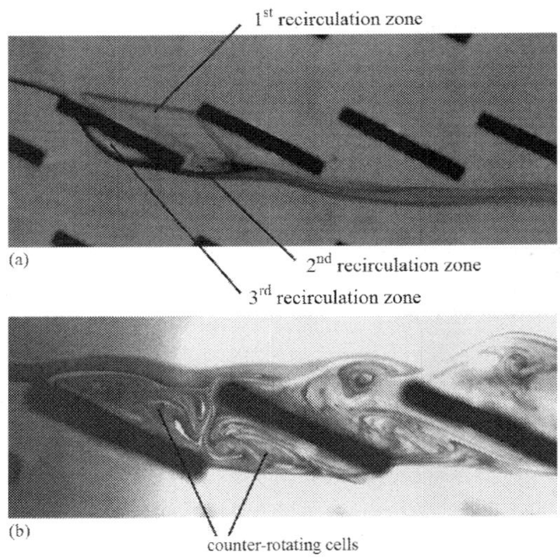

Figure 8. Flow structure around louvered fins: (a) Re = 510, (b) Re = 820 [8].

"third recirculation zone" forms on the upstream side of each louver. This third recirculation zone is caused when the flow passes through the gap between the first recirculation zone and the next louver downstream. This gap is narrow near the leading edge and much larger near the trailing edge. Flow in this gap must decelerate as it passes between the louvers. At all but the lowest Reynolds numbers where the first separation zone is small, the adverse pressure gradient in the inter-louver gap is large enough to cause separation from the upstream face of the next louver. The result is the third recirculation zone with counter-clockwise circulation [8].

Figure 8b shows flow through the same geometry at a higher Reynolds number (820) where the flow has become unsteady. The third recirculation zone has grown to become as large as the first zone, and the small second zone is no longer clearly evident. Two counter-rotating cells are present between the louvers. Fluid is periodically entrained in the recirculation zones and then ejected in the form of vortices which are carried downstream [8].

Today, the visualization techniques are developing parallel with the rapidly increasing technology. Besides the numerical analysis, infrared technology is a method to identify the flow characteristics of any flow. An open-circuit wind tunnel equipped with an infrared thermo-vision is illustrated in **Figure 9**. Infrared temperature measurement is achieved using an infrared camera. The electromagnetic energy radiated in the infrared spectral band by an object is transformed into an electronic signal by each of the thermo-vision sensors and is obtained simultaneously across the whole field of view, which depend on the optical focal length and the viewing distance [9].

In **Figure 10**, comparison of the experimental results obtained by the infrared measurements and the results of numerical analysis for the same louvered fin geometry is illustrated. The

1. Temperature controller
2. Anemometer
3. Pressure gauge
4. Surge tank
5. Flow control valve
6. Power supply
7. Infrared thermal imager
8. Test model
9. Air blower
10. Data acquisition instrument

Figure 9. Wind tunnel test equipped with an infrared thermo-vision [9].

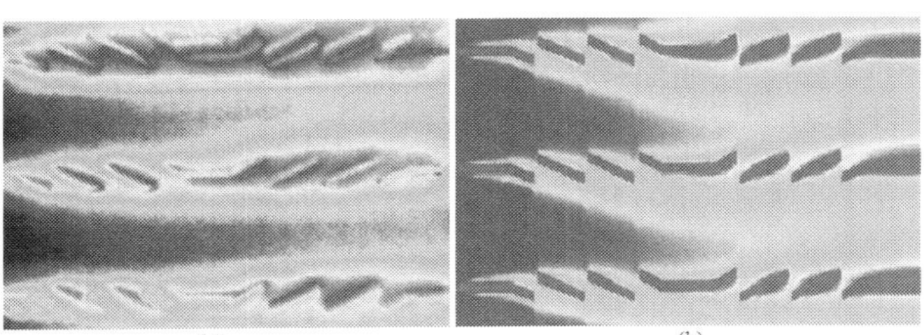

(a) (b)

Figure 10. Comparison of the temperature distribution of an infrared thermographs and numerical simulation results for a louvered fin array [9].

louvered fin has a louver angle of 20° and the flow velocity is 1.0 m/s. It is observed that both methods give similar temperature distributions across the entire louvered fin.

4. Data reduction in a louvered fin heat exchanger

In this section, the calculation method of the performance of a louvered fin heat exchanger is summarized. The equations will be written by considering the following assumptions.

I. Cold flow is the external flow of the louvered fin heat exchanger.

II. Hot flow is the internal flow of the louvered fin heat exchanger.

III. Thermophysical properties of both fluid are constant.

IV. The louvered fins are attached to a mini-channel flat tube.

V. Cold flow is uniform at the inlet of the louvered fin for the numerical analysis.

VI. The temperature of the tube wall is constant for the analytical solutions.

The main problem is the determination of external side heat transfer coefficient for the louvered fin surfaces by experimentally. Effectiveness-*NTU* method is generally used to determine the external side heat transfer coefficient by following the Kim and Bullard [10] method.

The average heat transfer rate can be expressed as

$$\dot{Q} = (\dot{Q}_c + \dot{Q}_h)/2 \tag{2}$$

where $\dot{Q}_c$ and $\dot{Q}_h$ are the heat transfer rates of cold and hot fluid, respectively. The heat transfer rates of each fluid can be calculated with Eqs. (3) and (4)

$$\dot{Q}_c = \dot{m}_c c_{p,c}(T_{c,o} - T_{c,i}) \tag{3}$$

$$\dot{Q}_h = \dot{m}_h c_{p,h}(T_{h,i} - T_{h,o}) \tag{4}$$

The effectiveness of the heat exchanger for one row configuration can be calculated using the following equation for both fluid unmixed

$$\varepsilon = 1 - exp\left[\frac{NTU^{0.22}}{C_r}\{\exp(-C_r NTU^{0.78}) - 1\}\right] \tag{5}$$

where

$$\varepsilon = \dot{Q}/\dot{Q}_{max} \tag{6}$$

$$C_r = \frac{(\dot{m}c_p)_{min}}{(\dot{m}c_p)_{max}} \tag{7}$$

We can obtain overall heat transfer coefficient (*UA*) for the heat exchanger as

$$UA = (\dot{m}c_p)_{min} NTU \tag{8}$$

where

$$UA = \dot{Q}/\Delta T_m \tag{9}$$

ΔT_m is the logarithmic mean temperature difference and that is

$$\Delta T_m = \frac{(T_{h,i}-T_{c,o})-(T_{h,o}-T_{c,i})}{\ln\left((T_{h,i}-T_{c,o})/(T_{h,o}-T_{c,i})\right)} \tag{10}$$

The external (cold) side heat transfer coefficient (h_c) can be obtained from the following equation by experimentally

$$\frac{1}{\eta_c A_c h_c} = \frac{1}{UA} - \frac{\delta_t}{k_t A_t} - \frac{1}{A_i h_i} \tag{11}$$

h_i is the internal side heat transfer coefficient, and it can be obtained using empirical relations for duct flow. The surface effectiveness (η_c) for a dry surface is

$$\eta_c = 1 - \frac{A_f}{A_c}(1-\eta_f) \tag{12}$$

where η_f is the efficiency of the fin as given in Eq. (13)

$$\eta_f = \frac{\tanh(ml)}{ml} \tag{13}$$

and

$$m = \sqrt{\frac{h_c P_f}{k_f A_{f,c}}} \tag{14}$$

Equation (14) can be expressed more explicitly using the following equation

$$m = \sqrt{\frac{2h_c}{k_f \delta}\left(1 + \frac{\delta}{F_d}\right)} \tag{15}$$

For the calculation of external side heat transfer coefficient (h_c) by analytically, the flow regime is very important to set up the analytical model. Therefore, Reynolds number is the mandatory parameter for the analytical solution. The flow can be assumed to be laminar at $Re_{L_p}<1300$ for the louvered fin arrays [3]. The characteristics length of louvered fin array is the louver pitch, so that the Reynolds number is calculated based on the louver pitch

$$Re_{L_p} = \frac{u_{max} L_p}{\upsilon} \tag{16}$$

u_{max} is the maximum velocity of the external fluid due to the narrowing section of the louvered fin arrays relatively to the inlet of the louvered fin. Therefore, the free velocity of the external fluid (u) at the inlet of the louvered fin is transformed to maximum velocity using the following equation

$$u_{max} = uF_p/(F_p - t) \tag{17}$$

where F_p and t are the fin pitch and the material thickness, respectively.

If the tube wall temperature is assumed constant for the numerical analysis, the heat transfer can be calculated by the heat gain of the external (cold) fluid as given in Eq. (3). Thus, the heat transfer coefficient of the external side (h_c) can be obtained from Eq. (18) by numerically

$$h_c = \dot{Q}_c/(A\Delta T_m) \tag{18}$$

In Eq. (18), A is the total external side heat transfer area and ΔT_m is the logarithmic mean temperature difference under constant wall temperature condition given by Eq. (19) [11]

$$\Delta T_m = \frac{(T_w - T_{c,o}) - (T_w - T_{c,i})}{\ln\left((T_w - T_{c,o})/(T_w - T_{c,i})\right)} \tag{19}$$

5. Performance evaluation criteria of the louvered fin heat exchangers

In the heat exchanger literature, some dimensionless parameters are used as a performance criteria. The commonly used thermal performance criteria are Stanton number and Colburn j-factor given as

$$St = \frac{h_c}{\rho u c_p} \tag{20}$$

$$j = St Pr^{2/3} \tag{21}$$

respectively. After the calculation of h_c by experimentally or analytically, Stanton number and Colburn j-factor can be obtained to indicate thermal performance of the louvered fin heat exchanger in a dimensionless form as given with Eqs. (20) and (21). Another performance criteria is the friction factor which is the dimensionless form of the pressure drop for the external side of a louvered fin heat exchanger expressed as

$$f = \left(\frac{A_c}{A}\right)\left(\frac{2\Delta P}{\rho u^2}\right) \tag{22}$$

An alternative equation for the friction factor can be used by considering the entrance, exit and acceleration effects

$$f = \left(\frac{A_c}{A}\right)\left(\frac{\rho_m}{\rho_1}\right)\left(\frac{2\rho_1 \Delta P}{G_c^2} - (k_c + 1 - \sigma^2) - 2\left(\frac{\rho_1}{\rho_2} - 1\right) + (1 - \sigma^2 - k_e)\frac{\rho_1}{\rho_2}\right) \tag{23}$$

where A_c is the minimum free flow area for the external side, and k_c and k_e are the coefficients of pressure loss at the inlet and the outlet of the heat exchanger. k_c and k_e can be evaluated according to Kays and London [2]. The overall performance of the louvered fin heat exchangers can be evaluated with another perspective. The ratio of the *j*-factor to the *f*, the ratio of the *j*-factor to the $f^{1/3}$ and *JF* are the overall performance criteria of the louvered fin heat exchangers used in the literature. j/f is known as "area goodness factor"[12, 13] and $j/f^{1/3}$ is known as "volume goodness factor"[6, 14, 15]. *JF* number which is related with the volume goodness factor can be obtained by Eq. (24) [16, 17]. These parameters are dimensionless numbers of the larger—the better characteristics. It is expected that these parameters can effectively evaluate the thermal and dynamic performance of a heat exchanger since it includes both the *j*- and the *f*-factor

$$JF = \frac{j/j_R}{(f/f_R)^{1/3}} \tag{24}$$

where j_R and f_R are the reference values of Colburn *j*-factor and friction factor, respectively.

In light of these explanations, thermal and hydraulic characteristics of the heat exchangers with louvered fins are presented using numerical and experimental studies in the literature.

In 1990s, 2D numerical models were preferred rather than 3D models due to the run time and limited computing power. Nevertheless, 3D models are necessary because of the high compatibility with the experimental results. The velocity and the temperature field of a louvered fin heat exchanger for two different Reynolds numbers are presented in **Figure 11** as a result of 2D numerical model.

It is observed that significant proportion of the air flows through the channels between the fins rather than between the louvers, as indicated by the presence of high velocity streaks in the channels at a Reynolds number of 100 (**Figure 11a, b**). The temperature of the air reach the fin temperature before it leaves the fin, therefore, the heat transfer performance of the second half of the fin is poor. In fact, second half of the fin only causes a pressure loss without any heat transfer at low Reynolds numbers. At a higher Reynolds number of 1600 (**Figure 11c, d**), the boundary layer of the louvers are much thinner, and therefore, the air is directed through the louver passages. A temperature difference is maintained between the air and the fin surface and so every part of the louvered fin contributes to the heat transfer. However, the 2D models is enough for the characteristics of the flow over the louvered fins, it is not possible to say same thing for the thermal performance. The comparison of 2D, 3D and the measured thermal and hydraulic performance of a louvered fin heat exchanger are presented in **Figure 12**.

It can be seen that the 2D model yields reasonably accurate predictions of friction factor, but poor predictions of Stanton number. An obvious way to identify the reasons of the error in the heat transfer is to consider the practical features which are missing from the 2-D model. Two important features which are missing are the tube surfaces and the resistance. The tube surfaces would add to the heat transfer area but would not add significantly to the overall heat transfer rate, because of the thick boundary layer growth on the tubes. The fin resistance would lower the temperature across the fin, and thus the heat transfer from the fin. Another reason of the over prediction of the thermal performance of the louvered fin is that the efficiency of the louvered fin cannot be calculated exactly. Generally, the experimental h_c value is obtained using plate fin surface of the fin efficiency even for the louver fin efficiency due to the absence of the base area of the fin in 2D models. In the literature, 2-D models only consider

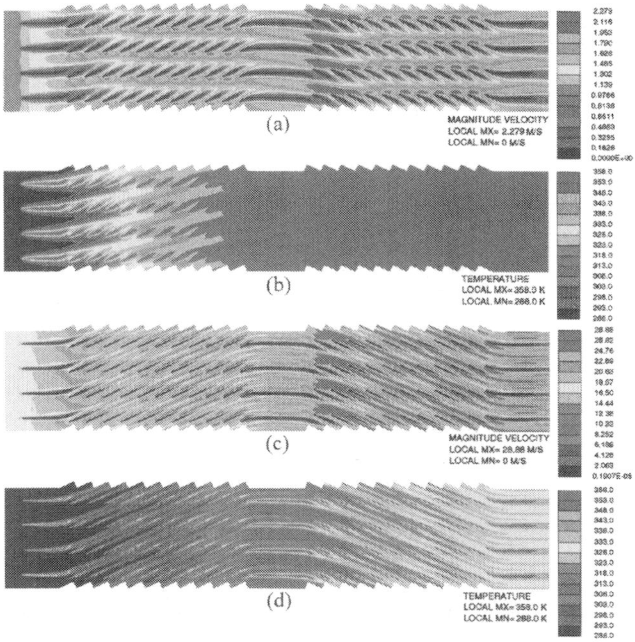

Figure 11. Computed velocity and temperature field for the 2-D model; $F_p = 2.54$ mm, $L_p = 1.4$ mm, $L_\alpha = 25.5°$. (a) Velocity, $Re_{Lp} = 100$, (b) Temperature, $Re_{Lp} = 100$, (c) Velocity, $Re_{Lp} = 1600$, (d) Temperature, $Re_{Lp} = 1600$ [18].

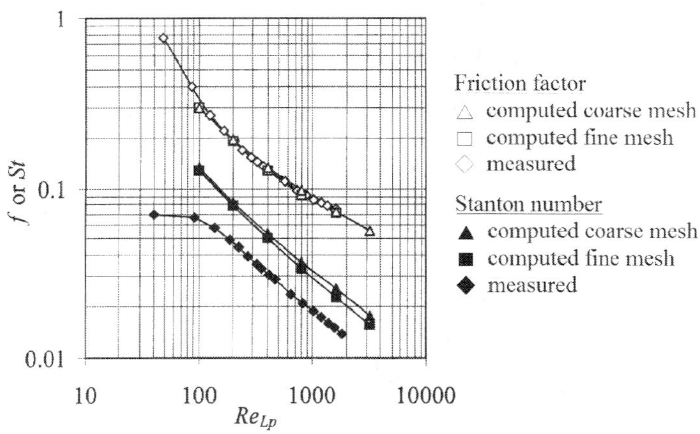

Figure 12. Comparison of the computed 2-D model and measured friction factor and Stanton number: $F_p = 2.05$ mm, $L_p = 1.4$ mm, $L_\alpha = 25.5°$, $T_p = 11$ mm and $\delta = 0.05$ mm [18].

the cross section of the louvers and the numerical h_c value is used directly without any fin efficiency formula. However, the slopes of the 2D numerical results are comparable and agree with the experimental results, heat transfer coefficient (h_c) is overpredicted with the assumption of constant fin temperature and neglecting the tube surface effect [6, 18, 19]. It is these factors which led to the development of the 3D models.

In 2000s, 3D models have come to the forefront with the increasing computing power. Researchers have begun to compare 2D and 3D results with their own experimental results to validate the compatibility of the numerical models. In **Figures 13** and **14**, the variation of Colburn *j*-factor and the friction factor with respect to the Reynolds number for a louvered fin heat exchanger is presented. The results are also compared with correlations in the literature. It is observed that the CFD results for the 2D models are overpredicted by 80% compared to the experimental results [19]. This is consistent with the study of Atkinson et al. [18]. The slopes of the experimental results are comparable and agree well with the correlated data. Colburn *j*-factor and the friction factor decrease with the increasing of Reynolds number.

Figures 12–14 demonstrate the compatibility of the numerical results with experimental results and the effect of Reynolds number on the thermal and hydraulic performance of the louvered fin heat exchanger. In addition to the effect of Reynolds number, the researchers spent great efforts to determine the optimum geometric parameters of the louvered fin. The numerical studies have a great importance in this field, because the testing of the every geometric variation is very difficult in terms of both time and cost. The prior geometric parameter is the louver angle which has the significant influence on the flow regime over the louvered fins. The variation of the thermal performance of a louvered fin heat exchanger for different louver pitch

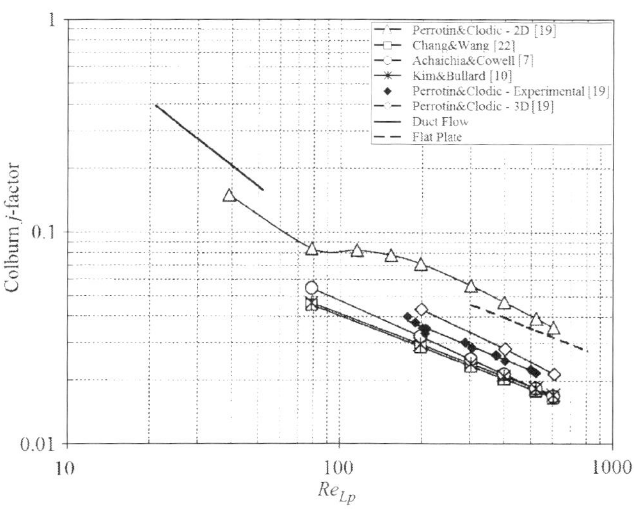

Figure 13. Comparison of computed and measured j-factor [19].

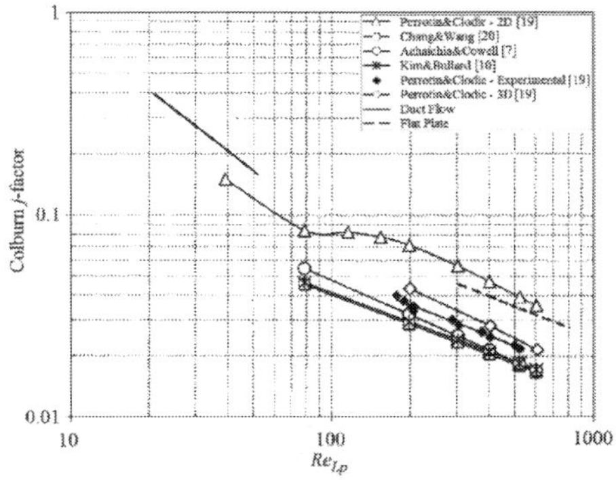

Figure 14. Comparison of computed and measured f-factor [19].

with respect to the louver angle is demonstrated in **Figure 15**. It is observed that the thermal performance is increasing up to the louver angle of 28.5° and then decreasing for all the louver pitches. The louver angle of 28.5° has the maximum thermal performance within the considered cases of the numerical study. Average heat transfer coefficient is about 200 W/m^{2}K at the minimum louver pitch of 0.81 mm. It decreases about to 185 W/m^{2}K at the maximum louver pitch of 1.4 mm [20].

In the study of Atkinson et al. [18], the louvered fin has uniform louver angle. It can be possible to create a louvered fin having non-uniform louver angles as shown in **Figure 16**. **Figure 17** shows the effect of non-uniform louver angle on the thermal and hydraulic performance of a louvered fin heat exchanger.

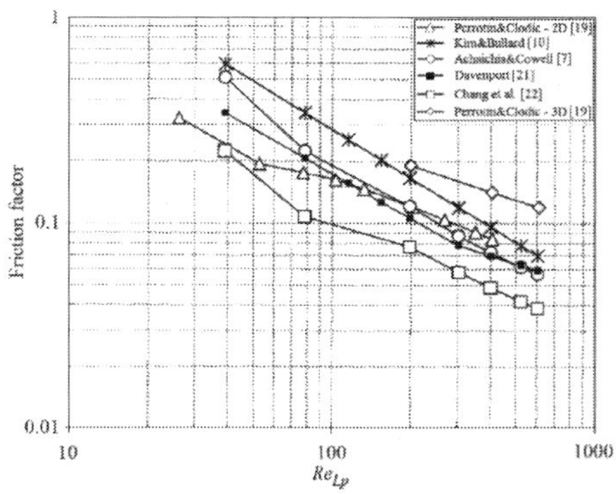

Figure 15. The effect of louver angle on the thermal performance of a louvered fin heat exchanger for different louver pitches [23].

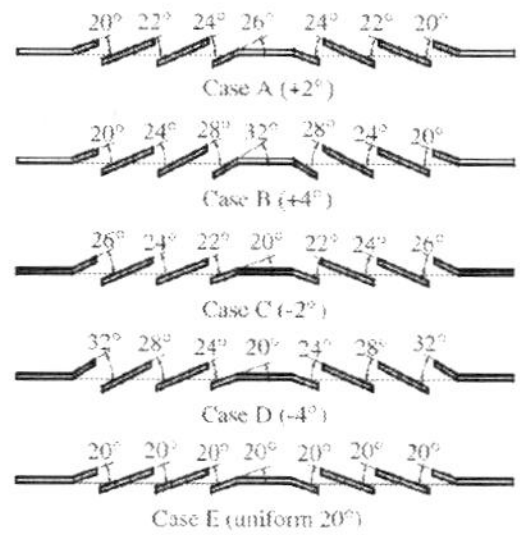

Figure 16. Five different cases of successively increased or decreased louver angle (+2°, +4°, −2°, −4°, and uniform angle 20°) [24].

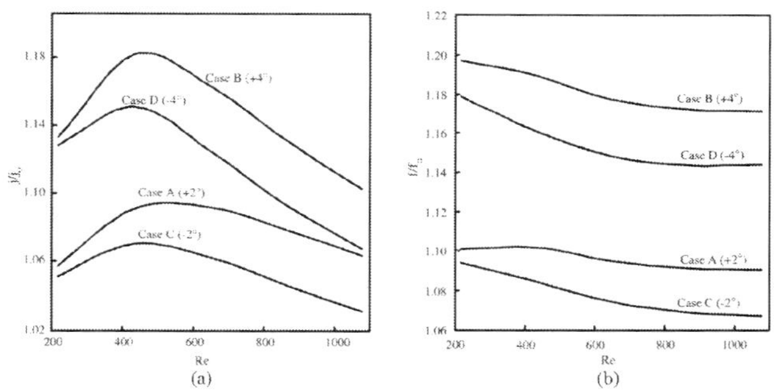

Figure 17. Effect of the non-uniform louver angle on the (a) Colburn j-factor and (b) friction factor [24].

In **Figure 17**, Colburn *j*-factor and the friction factor were normalized with Case E as shown in **Figure 16**. It is seen that the non-uniform louver angle patterns applied in the heat exchangers could effectively enhance the heat transfer performance. Case B has 18% heat transfer enhancement with respect to the Case E about at *Re* = 440. On the other hand, it has a negative effect of %19 on the friction factor.

In most cases, the geometric effects on the performance of a louvered fin heat exchanger are not monotonic. The combined relationship between the louver angle, louver pitch, fin pitch, tube pitch, etc., corrupts the linearity between the geometric parameters and the performance of the heat exchanger. An example is given with **Figure 18**. The variation of the heat transfer coefficient with respect to the frontal air velocity for different tube pitches is illustrated. It is observed that the heat transfer coefficient does not vary with the tube pitch linearly (**Figure 18a**). In **Figure 18b**, the effect of the fin pitch on the heat transfer coefficient is shown. It is seen that the heat transfer coefficient is increasing with decreasing of the fin pitch at a frontal velocity greater than 5.5 m/s. This statement is not valid for the frontal velocity smaller than 5.5 m/s.

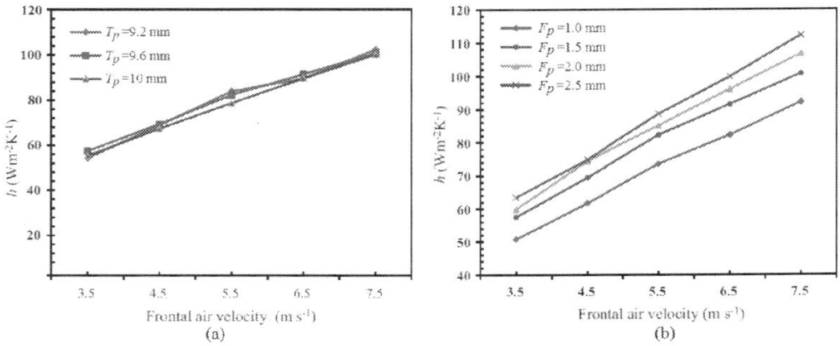

Figure 18. Variation of the heat transfer coefficient with respect to the frontal velocity for different (a) tube pitches (F_p = 1.5 mm, L_p = 1.2 mm, L_α = 26°) and (b) fin pitches (T_p = 9.6 mm, L_p = 1.2 mm, L_α = 26°) [25].

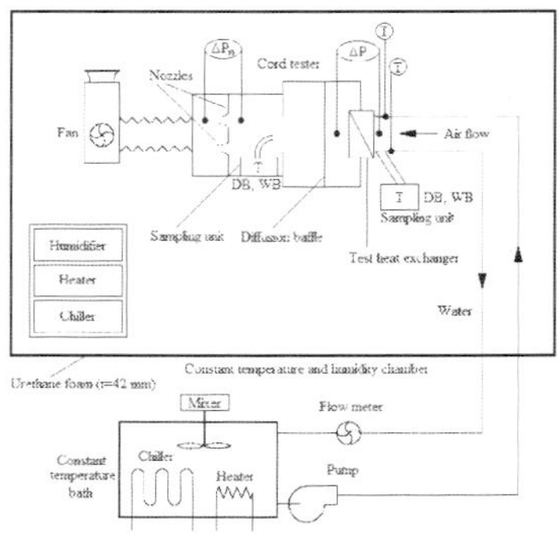

Figure 19. Schematic of a wind tunnel test [10].

Researchers make great efforts to identify the real performances of the louvered fin heat exchangers by experimentally. Investigation of the performance of the louvered fins is commonly performed with the wind tunnel tests. In the open literature, several wind tunnel test can be found in different designs. A typical wind tunnel is shown in **Figure 19**.

As shown in **Figure 19**, internal fluid of the heat exchanger is water and it is regulated by a constant temperature bath. External fluid of the heat exchanger is air, and it is sucked by a fan and wind tunnel is placed in a constant temperature and humidity chamber to regulate the air flow. Dry and wet bulb temperatures of the air are measured with thermocouples at the inlet and the exit of the heat exchanger. One of the most comprehensive performance data of the louvered fin heat exchangers is presented by this wind tunnel test in the early of 2000s. The effects of the geometric dimensions of the louvered fins and the Reynolds number on the Colburn *j*-factor and friction factor is identified. Similarly to the previous studies, *j*-factor and *f*-factor are decreasing with the increasing of Reynolds number due to its definition as shown in **Figure 20**.

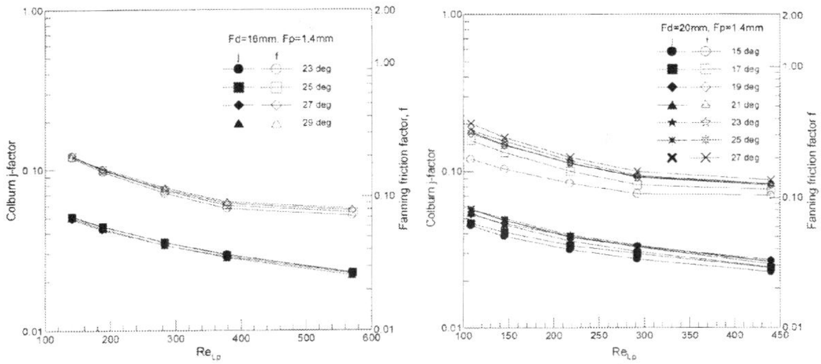

(a) (b)

Figure 20. Variations of j-factor and f-factor with respect to the louver angle and Reynolds number for (a) F_d = 16 mm and (b) F_d = 20 mm [10].

The effect of louver angle is more specifically at a flow depth of 20 mm for a constant fin pitch of 1.40 mm. Friction factor increases with the increasing of louver angle, however, the effect of the louver angle is diminishing for the louver angles larger than 21°. The effect of the louver angle on the *j*-factor varies with the flow depth. The increasing in the *j*-factor with the louver angle at a flow depth of 20 mm is more obvious than that of the flow depth of 16 mm. *j*-factor is increasing especially with the louver angle of greater than 23° due to the louver directed flow for such a small fin pitch of 1.40 mm.

The heat transfer and the pressure drop behaviour of the louvered fin heat exchangers for greater flow depths and Reynolds numbers is available in the open literature. The variations of the *j*-factor and *f*-factor with respect to the frontal air velocity for different geometric dimensions are given in **Figure 21**. As shown in **Figure 21**, the considered flow depth and Reynolds number range are 36.0–65.0 mm and 200–2500, respectively.

In **Figure 21a, c**, the effect of fin pitch at a constant flow depth and fin height on the thermal and hydraulic performance is shown. It is observed that the fin pitch has a significant effect on the thermal and hydraulic performance and *j*-factor and *f*-factor are decreasing with the increasing of fin pitch. For F_p = 2.0 mm and F_d = 65 mm, Colburn *j*-factor is maximum for all Reynolds numbers and it decreases about 0.0105–0.0072 as shown in **Figure 21a**. The cause of this situation is that the hydraulic resistance against the flow increases when the fin pitch decreases at the flow passage. Therefore, the flow tends to more to being louver directed. As a result of this phenomenon, the air flow can be mixed well by the louvers, so the heat transfer and the pressure drop increase. As shown in **Figure 21b**, *j*-factor and *f*-factor increase when the fin height increases. The possible reason is that the proportion of the air flow directed by the louvers increases with the fin height. **Figure 21d** shows the effect of the flow depth on the *j*-factor and *f*-factor. It is obvious that the flow depth has more significant effect on the *j*-factor and *f*-factor. The study of Kim and Bullard [10] indicates the same results. In addition to the *j*-factor and *f*-factor, the variation of volume goodness factor denoted by $j/f^{1/3}$ versus Reynolds number is illustrated in **Figure 22** as a performance criteria of the louvered fin heat exchangers.

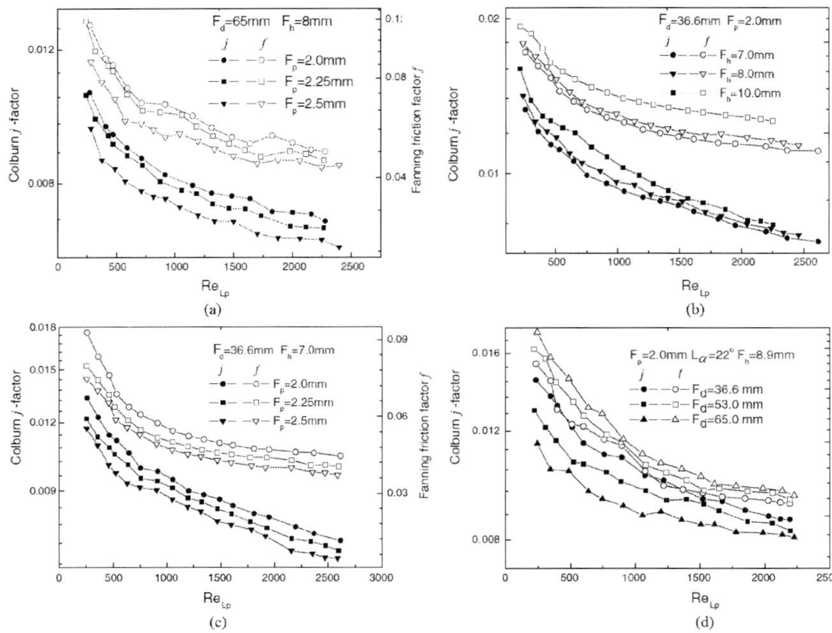

Figure 21. Variations of j-factor and f-factor with respect to the frontal air velocity [15].

It is seen that the geometry which has the smallest flow depth (36.0 mm), smallest fin pitch (2.00 mm) and the biggest fin height (10.0 mm) has the maximum value of $j/f^{1/3}$ = 0.032. The

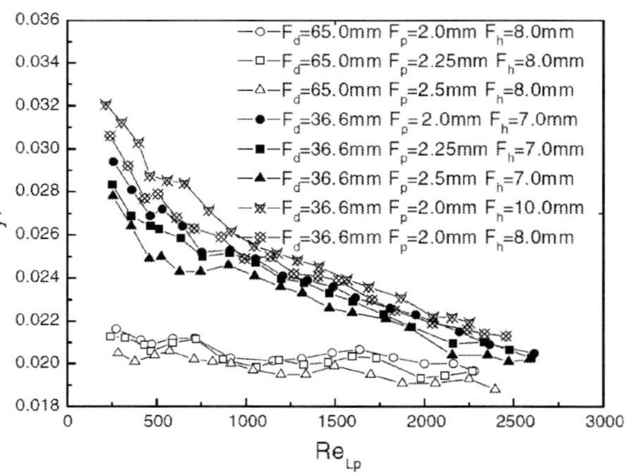

Figure 22. The variation of $j/f^{1/3}$ [15].

http://dx.doi.org/10.5772/66472

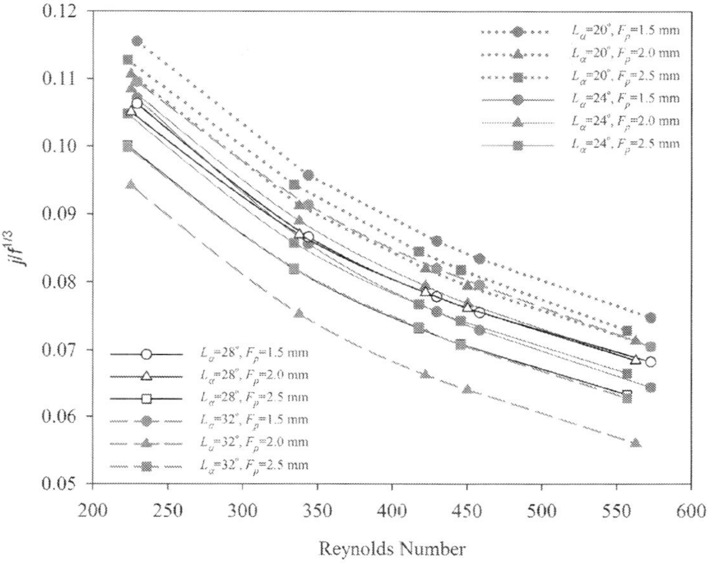

Figure 23. The variation of $j/f^{1/3}$ at low Reynolds number [6].

effect of the Reynolds number to the $j/f^{1/3}$ ratio decreases with a flow depth of 65.0 mm. The most important result is the non-monotonic behaviour of the $j/f^{1/3}$ ratio with the geometric dimensions. The variation of the $j/f^{1/3}$ ratio has a complicated behaviour with respect to the fin pitch and the fin height. In particular, the low Reynolds number region which has the significant changes of $j/f^{1/3}$ ratio for the smaller flow depths can be illuminated with another study by Erbay et al. [6] as shown in **Figure 23**.

The $j/f^{1/3}$ ratios of the study of Erbay et al. [6] are obtained for a constant flow depth of 20 mm by numerically. It seen that the $j/f^{1/3}$ ratio for all the geometries has a similar trend with respect to the Reynolds number; however, there is not linear relationship for the geometric parameters. As a result, such a performance criteria which considers both the thermal and hydraulic performance is necessary to design a heat exchanger.

Some of the researches working on the performance evaluation of the louvered fin heat exchangers have developed correlations for Stanton number, Colburn *j*-factor and friction factor. The basic form of these correlations is

$$a = C_1 Re^{C_2} \tag{25}$$

where a represents the performance criteria, and C_1 and C_2 are dependent on the dimensions of the louvered fin heat exchangers. Some of the correlations for the performance criteria of the louvered fin heat exchangers are listed by considering the studies in the open literature.

- Achaichia and Cowell [7], $150 < Re_{L_p} < 3000$

$$St = 1.54Re_{L_p}^{-0.57}\left(\frac{F_p}{L_p}\right)^{-0.19}\left(\frac{T_p}{L_p}\right)^{-0.11}\left(\frac{L_h}{L_p}\right)^{0.15} \tag{26}$$

$$f = 533.42Re_{L_p}^{0.318(logRe_{L_p}-2.25)}F_p^{-0.22}L_p^{0.25}T_p^{0.26}L_h^{0.33}. \tag{27}$$

- Kim and Bullard [10], $100 < Re_{L_p} < 600$

$$j = Re_{L_p}^{-0.487}\left(\frac{L_\alpha}{90}\right)^{0.257}\left(\frac{F_p}{L_p}\right)^{-0.13}\left(\frac{F_h}{L_p}\right)^{-0.29}\left(\frac{F_d}{L_p}\right)^{-0.235}\left(\frac{L_h}{L_p}\right)^{0.68}\left(\frac{T_p}{L_p}\right)^{-0.279}\left(\frac{\delta}{L_p}\right)^{-0.05} \tag{28}$$

$$f = Re_{L_p}^{-0.781}\left(\frac{L_\alpha}{90}\right)^{0.444}\left(\frac{F_p}{L_p}\right)^{-1.682}\left(\frac{F_h}{L_p}\right)^{-1.22}\left(\frac{F_d}{L_p}\right)^{0.818}\left(\frac{L_h}{L_p}\right)^{1.97}. \tag{29}$$

- Dong et al. [15], $200 < Re_{L_p} < 2500$

$$j = 0.26712Re_{L_p}^{-0.1944}\left(\frac{L_\alpha}{90}\right)^{0.257}\left(\frac{F_p}{L_p}\right)^{-0.5177}\left(\frac{F_h}{L_p}\right)^{-1.9045}\left(\frac{F_d}{L_p}\right)^{-0.2147}\left(\frac{L_h}{L_p}\right)^{1.7159}\left(\frac{\delta}{L_p}\right)^{-0.05} \tag{30}$$

$$f = 0.54486Re_{L_p}^{-0.3068}\left(\frac{L_\alpha}{90}\right)^{0.444}\left(\frac{F_p}{L_p}\right)^{-0.9925}\left(\frac{F_h}{L_p}\right)^{-0.5458}\left(\frac{F_d}{L_p}\right)^{-0.0688}\left(\frac{L_h}{L_p}\right)^{-0.2003} \tag{31}$$

- Ryu and Lee [26], $100 < Re_{L_p} < 3000$

$$j = Re_{L_p}^{(-0.484-1.887/lnRe_{L_p})}\left(\frac{F_d}{L_p}\right)^{0.157}\left(2.24-0.588ln\left(\frac{F_p}{L_p sinL_\alpha}\right)\right) \tag{32}$$

$$f = Re_{L_p}^{-0.433}\left(\frac{F_d}{L_p}\right)^{0.185}\left(1.10 + 4.31\left(\frac{L_\alpha}{90}\right)^2 + 0.836\frac{ln(F_p/L_p)}{(F_p/L_p)^2}\right). \tag{33}$$

In addition to above correlations, a large data bank as shown in **Table 1** was used by Chang and Wang [20], Chang et al. [22], and Park and Jacobi [27] to develop a more sensible correlations for *j*-factor and *f*-factor. These correlations which are the more comprehensive for the louvered fin heat exchangers are listed below.

- Chang and Wang [20], $100 < Re_{L_p} < 3000$

$$j = Re_{L_p}^{-0.49}\left(\frac{L_\alpha}{90}\right)^{0.27}\left(\frac{F_p}{L_p}\right)^{-0.14}\left(\frac{F_h}{L_p}\right)^{-0.29}\left(\frac{T_d}{L_p}\right)^{-0.23}\left(\frac{L_h}{L_p}\right)^{0.68}\left(\frac{T_p}{L_p}\right)^{-0.28}\left(\frac{\delta}{L_p}\right)^{-0.05}. \tag{34}$$

- Chang et al. [22], $100 < Re_{L_p} < 3000$

Source	L_p(mm)	F_p(mm)	F_h(mm)	L_h(mm)	L_α(°)	F_d(mm)	T_d(mm)	T_p(mm)	Δ(mm)	N_{LB}
D(1)[a, b] [21]	3.00	1.55	12.70	9.50	8.4	40.0	40.0	14.00	0.075	2
D(2)[a, b] [21]	3.00	1.55	12.70	9.50	10.4	40.0	40.0	14.00	0.075	2
D(3)[a, b] [21]	3.00	1.60	12.70	9.50	16.7	40.0	40.0	14.00	0.075	2
D(4)[a, b] [21]	2.25	1.55	12.70	9.50	13.4	40.0	40.0	14.00	0.075	2
D(5)[a, b] [21]	2.25	1.56	12.70	9.50	16.0	40.0	40.0	14.00	0.075	2
D(6)[a, b] [21]	2.25	1.56	12.70	9.50	19.2	40.0	40.0	14.00	0.075	2
D(7)[a, b] [21]	1.80	1.55	12.70	9.50	18.8	40.0	40.0	14.00	0.075	2
D(8)[a, b] [21]	1.80	1.59	12.70	9.50	20.8	40.0	40.0	14.00	0.075	2
D(9)[a, b] [21]	1.80	1.58	12.70	9.50	27.8	40.0	40.0	14.00	0.075	2
D(10)[a, b] [21]	1.50	1.53	12.70	9.50	19.6	40.0	40.0	14.00	0.075	2
D(11)[a, b] [21]	1.50	1.59	12.70	9.50	22.8	40.0	40.0	14.00	0.075	2
D(12)[a, b] [21]	1.50	1.60	12.70	9.50	35.9	40.0	40.0	14.00	0.075	2
D(13)[a, b] [21]	1.80	1.63	12.70	9.50	14.2	40.0	40.0	14.00	0.075	2
D(14)[a, b] [21]	3.00	1.56	12.70	9.50	11.2	40.0	40.0	14.00	0.075	2
D(15)[a, b] [21]	2.25	1.68	12.70	11.70	24.1	40.0	40.0	14.00	0.075	2
D(16)[a, b] [21]	2.25	1.65	12.70	11.00	21.4	40.0	40.0	14.00	0.075	2
D(17)[a, b] [21]	2.25	1.65	12.70	10.00	21.4	40.0	40.0	14.00	0.075	2
D(18)[a, b] [21]	2.25	1.63	12.70	9.00	21.4	40.0	40.0	14.00	0.075	2
D(19)[a, b] [21]	2.25	1.60	12.70	8.00	20.3	40.0	40.0	14.00	0.075	2
D(20)[a, b] [21]	3.00	1.54	7.80	7.10	13.9	40.0	40.0	9.18	0.075	2
D(21)[a, b] [21]	2.25	1.53	7.80	7.10	13.8	40.0	40.0	9.17	0.075	2
D(22)[a, b] [21]	1.80	1.54	7.80	7.10	20.4	40.0	40.0	9.18	0.075	2
D(23)[a, b] [21]	1.50	1.55	7.80	7.10	26.1	40.0	40.0	9.19	0.075	2
D(24)[a, b] [21]	2.25	1.49	7.80	7.10	9.5	40.0	40.0	9.14	0.075	2
D(25)[a, b] [21]	2.25	1.51	7.80	7.10	16.5	40.0	40.0	9.16	0.075	2
D(26)[a, b] [21]	2.25	1.51	7.80	7.10	17.7	40.0	40.0	9.16	0.075	2
D(27)[a, b] [21]	2.25	1.23	7.80	7.10	16.0	40.0	40.0	8.93	0.075	2
D(28)[a, b] [21]	2.25	1.01	7.80	7.10	13.9	40.0	40.0	8.74	0.075	2
D(29)[a, b] [21]	2.25	1.54	7.80	7.10	14.2	40.0	40.0	9.18	0.075	2
D(30)[a, b] [21]	2.25	1.51	7.80	6.50	16.6	40.0	40.0	9.16	0.075	2
D(31)[a, b] [21]	2.25	1.54	7.80	6.00	17.6	40.0	40.0	9.18	0.075	2
D(32)[a, b] [21]	2.25	1.50	7.80	5.00	14.4	40.0	40.0	9.15	0.075	2
CW(1)[a, b] [22]	1.32	1.80	16.00	12.44	28.0	22.0	22.0	21.00	0.160	2
CW(2)[a, b] [20]	1.32	2.00	16.00	12.44	28.0	22.0	22.0	21.00	0.160	2
CW(3)[a, b] [20]	1.32	2.20	16.00	12.44	28.0	22.0	22.0	21.00	0.160	2
CW(4)[a, b] [20]	1.42	1.80	19.00	17.18	28.0	22.0	22.0	24.00	0.160	2

Source	L_p(mm)	F_p(mm)	F_h(mm)	L_h(mm)	L_α(°)	F_d(mm)	T_d(mm)	T_p(mm)	Δ(mm)	N_{LB}
CW(5)[a, b] [20]	1.42	2.00	19.00	17.18	28.0	22.0	22.0	24.00	0.160	2
CW(6)[a, b] [20]	1.42	2.20	19.00	17.18	28.0	22.0	22.0	24.00	0.160	2
CW(7)[a, b] [20]	1.48	1.80	16.00	12.78	28.0	26.0	26.0	21.00	0.160	2
CW(8)[a, b] [20]	1.48	2.00	16.00	12.78	28.0	26.0	26.0	21.00	0.160	2
CW(9)[a, b] [20]	1.48	2.20	16.00	12.78	28.0	26.0	26.0	21.00	0.160	2
CW(10)[a, b] [20]	1.53	1.80	19.00	16.07	28.0	26.0	26.0	24.00	0.160	2
CW(11)[a, b] [20]	1.53	2.00	19.00	16.07	28.0	26.0	26.0	24.00	0.160	2
CW(12)[a, b] [20]	1.53	2.20	19.00	16.07	28.0	26.0	26.0	24.00	0.160	2
CW(13)[a, b] [20]	1.69	1.80	16.00	12.15	28.0	32.0	32.0	21.00	0.160	2
CW(14)[a, b] [20]	1.69	2.00	16.00	12.15	28.0	32.0	32.0	21.00	0.160	2
CW(15)[a, b] [20]	1.69	2.20	16.00	12.15	28.0	32.0	32.0	21.00	0.160	2
CW(16)[a, b] [20]	1.55	1.80	19.00	16.17	28.0	32.0	32.0	24.00	0.160	4
CW(17)[a, b] [20]	1.55	2.00	19.00	16.17	28.0	32.0	32.0	24.00	0.160	4
CW(18)[a, b] [20]	1.55	2.20	19.00	16.17	28.0	32.0	32.0	24.00	0.160	4
CW(19)[a, b] [20]	1.86	1.80	19.00	15.25	28.0	38.0	38.0	24.00	0.160	4
CW(20)[a, b] [20]	1.86	2.00	19.00	15.25	28.0	38.0	38.0	24.00	0.160	4
CW(21)[a, b] [20]	1.86	2.20	19.00	15.25	28.0	38.0	38.0	24.00	0.160	4
CW(22)[a, b] [20]	1.59	1.80	16.00	13.18	28.0	44.0	44.0	21.00	0.160	2
CW(23)[a, b] [20]	1.59	2.00	16.00	13.18	28.0	44.0	44.0	21.00	0.160	2
CW(24)[a, b] [20]	1.59	2.20	16.00	13.18	28.0	44.0	44.0	21.00	0.160	2
CW(25)[a, b] [20]	1.53	1.80	19.00	16.84	28.0	44.0	44.0	24.00	0.160	4
CW(26)[a, b] [20]	1.53	2.00	19.00	16.84	28.0	44.0	44.0	24.00	0.160	4
CW(27)[a, b] [20]	1.53	2.20	19.00	16.84	28.0	44.0	44.0	24.00	0.160	4
PSU(1)[a] [20][f]	1.00	1.124	8.00	6.50	30.0	16.0	16.0	9.60	0.157	2
PSU(2)[a] [20][f]	1.016	1.954	9.22	6.858	27.0	20.32	20.32	11.11	0.0508	2
PSU(3)[a] [20][f]	1.016	1.588	9.22	6.858	27.0	20.32	20.32	11.11	0.0508	2
PSU(4)[a] [20][f]	1.016	1.270	9.22	6.858	27.0	20.32	20.32	11.11	0.0508	2
PSU(5)[a] [20][f]	0.94	1.114	9.15	7.62	27.0	16.26	16.26	11.11	0.127	2
AC(1)[a, b] [7]	1.40	2.02	9.00	8.50	25.5	41.6	32.0	11.00	0.05	2
AC(2)[a, b] [7]	1.40	3.25	9.00	8.50	25.5	41.6	32.0	11.00	0.05	2
AC(3)[a, b] [7]	1.40	1.65	9.00	8.50	25.5	41.6	32.0	11.00	0.05	2
AC(4)[a, b] [7]	1.40	2.09	9.00	8.50	21.5	41.6	32.0	11.00	0.05	2
AC(5)[a, b] [7]	1.40	2.03	9.00	8.50	28.5	41.6	32.0	11.00	0.05	2
AC(6)[a, b] [7]	1.40	2.15	9.00	8.50	25.5	20.8	16.0	11.00	0.05	1
AC(7)[a, b] [7]	1.40	1.70	9.00	8.50	25.5	20.8	16.0	11.00	0.05	1
AC(8)[a, b] [7]	0.81	2.11	9.00	8.50	29.0	41.6	32.0	11.00	0.05	2

Source	L_p(mm)	F_p(mm)	F_h(mm)	L_h(mm)	L_α(°)	F_d(mm)	T_d(mm)	T_p(mm)	Δ(mm)	N_{LB}
AC(9)[a, b] [7]	0.81	1.72	9.00	8.50	29.0	41.6	32.0	11.00	0.05	2
AC(10)[a, b] [7]	0.81	3.33	9.00	8.50	29.0	41.6	32.0	11.00	0.05	2
AC(11)[a, b] [7]	1.10	2.18	9.00	8.50	30.0	41.6	32.0	11.00	0.05	2
AC(12)[a, b] [7]	0.81	2.16	9.00	8.50	20.0	41.6	32.0	11.00	0.05	2
AC(13)[a, b] [7]	1.10	2.16	6.00	5.50	28.0	41.6	32.0	8.00	0.05	2
AC(14)[a, b] [7]	1.10	2.17	12.00	11.50	22.0	41.6	32.0	14.00	0.05	2
AC(15)[a, b] [7]	1.10	2.17	9.00	5.50	22.0	41.6	32.0	8.00	0.05	2
WJ(1)[a, b] [28]	1.397	2.117	18.923	16.255	30.0	25.4	25.4	22.99	0.158	2
WJ(2)[a, b] [28]	1.397	1.693	18.923	16.255	30.0	25.4	25.4	22.99	0.158	2
WJ(3)[a, b] [28]	1.397	1.411	18.923	16.255	30.0	25.4	25.4	22.99	0.158	2
WJ(4)[a, b] [28]	1.65	2.117	8.64	7.0987	30.0	25.4	25.4	22.99[c]	0.158	2
WJ(5)[a, b] [28]	1.65	1.693	8.64	7.0987	30.0	25.4	25.4	22.99[c]	0.158	2
WJ(6)[a, b] [28]	1.65	1.411	8.64	7.0987	30.0	25.4	25.4	22.99[c]	0.158	2
R(1)[a, b] [29]	0.85	0.51	2.84	2.13	25.0	15.6	15.6	7.51[c]	0.025	2
SS(1)[a, b] [30]	1.40	1.50	12.5	10.20	22.0	57.4	57.4	14.00[d]	0.06	3
SS(2)[a, b] [30]	1.40	2.00	12.4	10.30	18.5	57.4	57.4	13.90[d]	0.06	2
SS(3)[a, b] [30]	1.30	2.00	12.4	10.00	24.5	37.0	37.0	13.90[d]	0.06	2
SS(4)[a, b] [30]	1.20	1.80	8.6	6.80	24.0	37.0	37.0	10.10[d]	0.04	2
SS(5)[a, b] [30]	1.10	1.80	9.6	6.80	25.5	50.0	50.0	11.10[d]	0.06	2
SS(6)[a, b] [30]	0.50	1.90	8.0	5.00	28.5	47.8	47.8	9.50[d]	0.04	4
T(1)[a] [31]	1.884	1.50	20.0	18.50	35.0	50.0	50.0	25[e]	0.16	2
J(1)[b] [32]	1.40	1.06	7.93	6.93	27.0	15.9	–	9.86	0.102	2
J(2)[b] [32]	1.40	2.12	7.93	6.93	27.0	27.9	–	9.86	0.102	2
J(3)[b] [32]	1.40	1.06	7.93	6.93	27.0	27.9	–	9.86	0.102	2
J(4)[b] [32]	1.14	5.08	12.43	11.15	29.0	25.4	–	14.26	0.114	2
J(5)[b] [32]	1.14	2.12	12.43	11.15	29.0	25.4	–	14.26	0.114	2
J(6)[b] [32]	1.14	1.41	12.43	11.15	29.0	25.4	–	14.26	0.114	2
KB(1)[b] [10]	1.70	1.40	8.15	6.40	15.0	20.0	–	10.15	0.10	2
KB(2)[b] [10]	1.70	1.40	8.15	6.40	17.0	20.0	–	10.15	0.10	2
KB(3)[b] [10]	1.70	1.40	8.15	6.40	19.0	20.0	–	10.15	0.10	2
KB(4)[b] [10]	1.70	1.40	8.15	6.40	21.0	20.0	–	10.15	0.10	2
KB(5)[b] [10]	1.70	1.40	8.15	6.40	23.0	20.0	–	10.15	0.10	2
KB(6)[b] [10]	1.70	1.40	8.15	6.40	25.0	20.0	–	10.15	0.10	2
KB(7)[b] [10]	1.70	1.40	8.15	6.40	27.0	20.0	–	10.15	0.10	2
KB(8)[b] [10]	1.70	1.40	8.15	6.40	23.0	24.0	–	10.15	0.10	2
KB(9)[b] [10]	1.70	1.40	8.15	6.40	25.0	24.0	–	10.15	0.10	2

Source	L_p(mm)	F_p(mm)	F_h(mm)	L_h(mm)	L_α(°)	F_d(mm)	T_d(mm)	T_p(mm)	Δ(mm)	N_{LB}
KB(10)[b] [10]	1.70	1.40	8.15	6.40	27.0	24.0	–	10.15	0.10	2
KB(11)[b] [10]	1.70	1.40	8.15	6.40	29.0	24.0	–	10.15	0.10	2
KB(12)[b] [10]	1.70	1.10	8.15	6.40	23.0	20.0	–	10.15	0.10	2
KB(13)[b] [10]	1.70	1.10	8.15	6.40	23.0	24.0	–	10.15	0.10	2
KB(14)[b] [10]	1.70	1.20	8.15	6.40	23.0	20.0	–	10.15	0.10	2
KB(15)[b] [10]	1.70	1.20	8.15	6.40	23.0	24.0	–	10.15	0.10	2
KYL(1)[b] [14]	2.90	2.82	16.50	12.50	20.0	44.0	–	21.20	0.15	2
KYL(2)[b] [14]	2.90	2.42	16.50	12.50	20.0	44.0	–	21.20	0.15	2
KYL(3)[b] [14]	2.90	2.03	16.50	12.50	20.0	44.0	–	21.20	0.15	2
KYL(4)[b] [14]	2.90	2.82	16.50	12.50	25.0	44.0	–	21.20	0.15	2
KYL(5)[b] [14]	2.90	2.42	16.50	12.50	25.0	44.0	–	21.20	0.15	2
KYL(6)[b] [14]	2.90	2.03	16.50	12.50	25.0	44.0	–	21.20	0.15	2
KYL(7)[b] [14]	2.90	2.82	16.50	12.50	30.0	44.0	–	21.20	0.15	2
KYL(8)[b] [14]	2.90	2.42	16.50	12.50	30.0	44.0	–	21.20	0.15	2
KYL(9)[b] [14]	2.90	2.03	16.50	12.50	30.0	44.0	–	21.20	0.15	2
KYL(10)[b] [14]	2.90	2.82	16.50	12.50	35.0	44.0	–	21.20	0.15	2
KYL(11)[b] [14]	2.90	2.42	16.50	12.50	35.0	44.0	–	21.20	0.15	2
KYL(12)[b] [14]	2.90	2.03	16.50	12.50	35.0	44.0	–	21.20	0.15	2

[a] Correlated data considered by Chang and Wang [20].[b] Correlated data considered by Park and Jacobi [27].[c] Park and Jacobi [27] used half of these pitches because of two fin stocks and a splitter plate between the tubes.[d] Park and Jacobi [27] and Chang and Wang [20] assumed that the tube diameter is 1.50 mm due to the lack of information.[e] Chang and Wang [20] assumed that the tube diameter is 5.00 mm due to the lack of information.[f] Correlated data which was unpublished by R. L. Webb considered by Chang and Wang [20].

Table 1. Geometrical dimensions of the louvered fin heat exchangers in the database [20, 27].

$$f = f_1 f_2 f_3 \tag{35}$$

for Re_{L_p}<150

$$f_1 = 14.39 Re_{L_p}^{\left(-0.805\frac{F_p}{F_h}\right)} \left(log_e\left(1.0 + \left(\frac{F_p}{L_p}\right)\right)\right)^{3.04} \tag{36}$$

$$f_2 = \left(log_e\left(\left(\frac{\delta}{F_p}\right)^{0.48} + 0.9\right)\right)^{-1.435} \left(\frac{D_h}{L_p}\right)\left(log_e(0.5Re_{L_p})\right)^{-3.01} \tag{37}$$

$$f_3 = \left(\frac{F_p}{L_h}\right)^{-0.308}\left(\frac{F_d}{L_h}\right)^{-0.308}\left(e^{-0.1167\frac{T_p}{D_m}}\right)L_\alpha^{0.35} \tag{38}$$

for $150 < Re_{L_p} < 5000$

$$f_1 = 4.97Re_{L_p}^{0.6049-1.064/L_\alpha^{0.2}}\left(log_e\left(\left(\frac{\delta}{F_p}\right)^{0.5}+0.9\right)\right)^{-0.527} \tag{39}$$

$$f_2 = \left(\left(\frac{D_h}{L_p}\right)log_e(0.3Re_{L_p})\right)^{-2.966}\left(\frac{F_p}{L_h}\right)^{-0.7931\frac{T_p}{T_h}} \tag{40}$$

$$f_3 = \left(\frac{T_p}{D_m}\right)^{-0.0446}log_e\left(1.2+\left(\frac{L_p}{F_p}\right)^{1.4}\right)^{-3.553}L_\alpha^{-0.477}. \tag{41}$$

- Park and Jacobi [27], $27 < Re_{L_p} < 4132$

$$j_{cor} = 0.872j_{Re}j_{low}j_{louver}L_\alpha^{0.219}N_{LB}^{-0.0881}\left(\frac{F_h}{L_p}\right)^{0.149}\left(\frac{F_d}{F_p}\right)^{-0.259}\left(\frac{L_h}{F_h}\right)^{0.54} \times\left(\frac{F_h}{T_p}\right)^{-0.902}\left(1-\frac{\delta}{L_p}\right)^{2.62}\left(\frac{L_p}{F_p}\right)^{0.301} \tag{42}$$

where

$$j_{Re} = Re_{L_p}^{[-0.458-0.00874\cosh(F_p/L_p-1)]} \tag{43}$$

$$j_{low} = 1-\sin\left(\frac{L_p}{F_p}L_\alpha\right)\left[\cos h\left(0.049Re_{L_p}-0.142\frac{F_d}{N_{LB}}\right)\right]^{-1} \tag{44}$$

$$j_{louver} = 1-(-0.0065\tan L_\alpha)\left(\frac{F_d}{N_{LB}F_p}\right)\cos\left[2\pi\left(\frac{F_p}{L_p\tan L_\alpha}-1.8\right)\right] \tag{45}$$

$$f_{cor} = 3.69f_{Re}N_{LB}^{-0.256}\left(\frac{F_p}{L_p}\right)^{0.904}\sin(L_\alpha+0.2)\left(1-\frac{F_h}{T_p}\right)^{0.733}\times\left(\frac{L_h}{F_h}\right)^{0.648}\left(\frac{\delta}{L_p}\right)^{-0.647}\left(\frac{F_h}{F_p}\right)^{0.799} \tag{46}$$

where

$$f_{Re} = \left(Re_{L_p}\frac{F_p}{L_p}\right)^{-0.845}+0.0013Re_{L_p}^{[1.26(\delta/F_p)]} \tag{47}$$

However, the correlations of Stanton number, Colburn *j*-factor and friction factor are defined for a large range of Reynolds number and geometric descriptions for heat exchangers with multi-louvered fins, it is necessary that the performance of every new type of heat exchanger is analysed individually due to the complexity of the combined effects of geometrical and operational parameters [33].

6. Concluding remarks

In this chapter, the structure of the louvered fin is examined in terms of thermal and hydraulic performance by following the studies in the literature. Several experimental and numerical studies are analysed by the authors to present a guide for the louvered fin. It is clear that the geometric parameters such as fin pitch, fin height, louver pitch, louver angle, and flow depth have remarkable effect on the performance of a louvered fin heat exchanger. It can be stated that the combined effects of these parameters must be examined individually to design a high efficiency heat exchanger. The key points of this chapter for the researchers can be summarized as follows:

- Frontal air velocity and the louver angle are the determinative parameters for the flow regime over the louvered fins. The duct directed or louver directed flow can be formed by the effects of these parameters.
- The flow efficiency over the louvered fins is increasing with the louver directed flow.
- 2D numerical models are inadequate to predict the heat transfer coefficient due to the lack of the un-finned areas. 3D numerical is sensible to predict both heat transfer coefficient and friction factor.
- Colburn *j*-factor increases with decreasing fin pitch.
- Friction factor (*f*) decreases with increasing fin pitch.
- The area goodness factor (j/f) and the volume goodness factor ($j/f^{1/3}$) decrease with increasing Reynolds number but they do not have a monotonic relation with the geometric parameters such as fin pitch and louver angle.
- However, the wide range of correlated data is available in the literature, every heat exchanger must be analysed individually due to combined effects of geometric parameters and operating conditions.

Acknowledgements

This research was performed under the Santez Project (00865-STZ.2011–1). The authors would like to thank the Ministry of Science, Industry and Technology and the Research and Development Department of the Arçelik A.Ş. Eskişehir Refrigerator Plant for their support.

Nomenclature

a	Performance criteria
A	Heat transfer area, m^2
A_c	Heat transfer area of the cold side, m^2
$A_{c,f}$	Cross-sectional area of the fin, m^2
A_f	Frontal area, m^2
A_i	Heat transfer area of the hot (internal) side, m^2
A_t	Heat transfer area of the tube, m^2
$c_{p,c}$	Specific heat of the cold fluid, kJ/(kg°C)
$c_{p,h}$	Specific heat of the hot fluid, kJ/(kg°C)
C_r	Heat capacity ratio
D	Ideal transverse distance, mm
D_h	Hydraulic diameter, mm
D_m	Major tube diameter, mm
f	Fanning friction factor
f_{cor}	Fanning friction factor correlation
F_d	Flow depth, mm
F_h	Fin height, mm
F_p	Fin pitch, mm
f_R	Reference value of the friction factor
f_{Re}	Correlation factor for Reynolds number effect
G_c	Air mass flux at minimum cross-sectional area, kg/(m^2s)
h	Heat transfer coefficient, W/(m^2°C)
h_c	Heat transfer coefficient of the cold (external) fluid, W/(m^2°C)
h_i	Heat transfer coefficient of the hot (internal) fluid, W/(m^2°C)
j	Colburn j-factor
j_{cor}	Colburn j-factor correlation
j_{louver}	Correlation factor for louver geometry effect
j_{low}	Correlation factor for low Reynolds number effect
j_R	Reference value of the Colburn j-factor
JF	Performance evaluation criteria related with volume goodness factor
k_c	Pressure loss coefficient at the inlet of the heat exchanger
k_e	Pressure loss coefficient at the exit of the heat exchanger
k_f	Thermal conductivity of the fin material, W/(m°C)
k_t	Thermal conductivity of the tube material, W/(m°C)
l	Half of the fin height (F_h), mm
L_α	Louver angle, degree
L_h	Louver height, mm
L_p	Louver pitch, mm
m	Parameter for the calculation of η_f related with h_c, $A_{c,f}$, P_f and k_f, m^{-1}
$\dot{m}_c$	Mass flow rate of the cold fluid, kg/s

$\dot{m}_h$	Mass flow rate of the hot fluid, kg/s
N	Actual transverse distance, mm
NTU	Number of transfer unit
P	Pressure, Pa
P_f	Perimeter of the fin, m
Pr	Prandtl number
$\dot{Q}$	Average heat transfer rate, W
$\dot{Q}_c$	Heat transfer rate of the cold fluid, W
$\dot{Q}_h$	Heat transfer rate of the hot fluid, W
$\dot{Q}_{max}$	Maximum heat transfer rate, W
Re	Reynolds number
Re_{L_p}	Reynolds number based on the louver pitch
T	Temperature, °C
$T_{c,i}$	Inlet temperature of the cold fluid, °C
$T_{c,o}$	Outlet temperature of the cold fluid, °C
T_h	Parameter with the tube pitch and the major tube diameter: T_p–D_m, mm
$T_{h,i}$	Inlet temperature of the hot fluid, °C
$T_{h,o}$	Outlet temperature of the hot fluid, °C
T_p	Tube pitch, mm
T_w	Wall temperature, °C
ΔT_m	Logarithmic mean temperature difference, °C
St	Stanton number
u	Free velocity of the external fluid, m/s
u_{max}	Maximum velocity of the external fluid, m/s
U	Overall heat transfer coefficient, W/(m^2°C)
UA	Overall thermal conductance, W/°C

Greek letters

δ	Thickness of the fin material, mm
δ_t	Thickness of the flat tube material, mm
ε	Effectiveness
η	Flow efficiency
η_c	Surface effectiveness of the cold side
η_f	Fin efficiency
υ	Kinematic viscosity, m^2/s
ρ	Density, kg/m^3

Subscripts

1 Inlet
2 Outlet
m Mean

Author details

Latife Berrin Erbay[1]*, Bahadır Doğan[1] and Mehmet Mete Öztürk[2]

*Address all correspondence to: lberbay@ogu.edu.tr

1 Department of Mechanical Engineering, Faculty of Engineering and Architecture, Eskişehir Osmangazi University, Eskişehir, Turkey

2 Department of Motor Vehicles and Transportation Technology, Vocational School of Transportation, Anadolu University, Eskişehir, Turkey

References

[1] Shah R K, Sekulic D P. Fundamental of Heat Exchanger Design. 1st ed. New Jersey: John Wiley & Sons, Inc.; 2003. 941 p.

[2] Kays W M, London A L. Compact Heat Exchangers. 3rd ed. New York: McGraw-Hill; 1984. 335 p.

[3] Webb R L, Trauger P. How structure in the louvered fin heat exchanger geometry. Experimental Thermal and Fluid Science. 1991;**4**(2):205–217. DOI: 10.1016/0894-1777(91)90065-Y

[4] Aoki H, Shinagawa T, Suga K. An experimental study of the local heat transfer characteristics in automotive louvered fins. Experimental Thermal and Fluid Science. 1989;2(3):293–300. DOI: 10.1016/0894-1777(89)90018-6

[5] Cowell T A, Heikal M R, Achaichia A. Flow and heat transfer in compact louvered fin surfaces. Experimental Thermal and Fluid Science. 1995;**10**(2):192–199. DOI: 10.1016/0894-1777(94)00093-N

[6] Erbay L B, Uğurlubilek N, Altun Ö, Dogan B. Numerical investigation of the air-side thermal hydraulic performance of a louvered-fin and flat-tube heat exchanger at low Reynolds numbers. Heat Transfer Engineering. 2017;**38**(5). DOI: 10.1080/01457632.2016.1200382

[7] Achaichia A, Cowell T A. Heat transfer and pressure drop characteristics of flat tube and louvered plate fin surfaces. Experimental Thermal and Fluid Science. 1988;**1**(2):147–157. DOI: 10.1016/0894-1777(88)90032-5

[8] DeJong N C, Jacobi A M. Flow, heat transfer, and pressure drop in the near-wall region of louvered-fin arrays. Experimental Thermal and Fluid Science. 2003;**27**(3):237–250. DOI: 10.1016/S0894-1777(02)00224-8

[9] Jang J-Y, Chen C-C. Optimization of louvered-fin heat exchanger with variable louver angles. Applied Thermal Engineering. 2015;**91**:138–150. DOI: 10.1016/j.applthermaleng.2015.08.009

[10] Kim M-H, Bullard C W. Air-side thermal hydraulic performance of multi-louvered fin aluminium heat exchangers. International Journal of Refrigeration. 2002;**25**(3):390–400. DOI: 10.1016/S0140-7007(01)00025-1

[11] Xiaoping T, Huahe L, Xaingfei T. CFD simulation and experimental study on airside performance for MCHX. In: International Refrigeration and Air Conditioning Conference; 12–15 July 2010; Purdue University, School of Mechanical Engineering; 2010. p. 1–8.

[12] Joardar A, Jacobi A M. Impact of leading edge delta-wing vortex generators on the thermal performance of a flat tube, louvered-fin compact heat exchanger. International Journal of Heat and Mass Transfer. 2005;**48**(8):1480–1493. DOI: 10.1016/j.ijheatmasstransfer.2004.10.018

[13] Han H, He Y-L, Li Y-S, Wang Y, Wu M. A numerical study on compact enhanced fin-and-tube heat exchangers exchangers with oval and circular tube configurations. International Journal of Heat and Mass Transfer. 2013;**65**:686–695. DOI: 10.1016/j.ijheatmasstransfer.2013.06.049

[14] Kim J H, Yun J H, Lee C S. Heat-transfer and friction characteristics for the louver-fin heat exchanger. Journal of Thermophysics and Heat Transfer. 2004;**18**(1):58–64. DOI: 10.2514/1.9123

[15] Dong J, Chen J, Chen Z, Zhang W, Zhou Y. Heat transfer and pressure drop correlations for the multi-louvered fin compact heat exchangers. Energy Conversion and Management. 2007;**48**(5):1506–1515. DOI: 10.1016/j.enconman.2006.11.023

[16] Yun J-Y, Lee K-S. Influence of design parameters on the heat transfer and flow friction characteristics of the heat exchanger with slit fins. International Journal of Heat and Mass Transfer. 2000;**43**(14):2529–2539. DOI: 10.1016/S0017-9310(99)00342–7

[17] Qi Z-G, Chen J-P, Chen Z-J. Parametric study on the performance of a heat exchanger with corrugated louvered fins. Applied Thermal Engineering. 2007;**27**(2–3):539–544. DOI: 10.1016/j.applthermaleng.2006.06.015

[18] Atkinson K N, Drakulic R, Heikal M R, Cowell T A. Two- and three-dimensional numerical models of flow and heat transfer over louvred fin arrays in compact heat exchangers. International Journal of Heat and Mass Transfer. 1998;**41**(24):4063–4080. DOI: 10.1016/S0017-9310(98)00165-3

[19] Perrotin T, Clodic D. Thermal-hydraulic CFD study in louvered fin-and-flat-tube heat exchangers. International Journal of Refrigeration. 2004;**27**(4):422–432. DOI: 10.1016/j.ijrefrig.2003.11.005

[20] Malapure V P, Mitra S K, Bhattacharya A. Numerical investigation of fluid flow and heat transfer over louvered fins in compact heat exchanger. International Journal of Thermal Sciences. 2007;**46**(2):199–211. DOI: 10.1016/j.ijthermalsci.2006.04.010

[21] Hsieh C-T, Jang J-Y. 3-D thermal-hydraulic analysis for louver fin heat exchangers with variable louver angle. Applied Thermal Engineering. 2006;**26**(14–15):1629–1639. DOI: 10.1016/j.applthermaleng.2005.11.019

[22] Chang Y-J, Wang C-C. A generalized heat transfer correlation for Iouver fin geometry. International Journal of Heat and Mass Transfer. 1997;**40**(3):533–544. DOI: 10.1016/0017-9310(96)00116–0

[23] Davenport C J. Correlations for heat transfer and flow friction characteristics of louvered fin. In: 21st National Heat Transfer Conference; Seattle, New York, USA: AIChE; 1983. p. 19–27.

[24] Chang Y-J, Hsu K-C, Lin Y-T, Wang C-C. A generalized friction correlation for louver fin geometry. International Journal of Heat and Mass Transfer. 2000;**43**(12):2237–2243. DOI: 10.1016/S0017-9310(99)00289-6

[25] Karthik P, Kumaresan V, Velraj R. Experimental and parametric studies of a louvered fin and flat tube compact heat exchanger using computational fluid dynamics. Alexandria Engineering Journal. 2015;**54**(4):905–915. DOI: 10.1016/j.aej.2015.08.003

[26] Ryu K, Lee K-S. Generalized heat-transfer and fluid-flow correlations for corrugated louvered fins. International Journal of Heat and Mass Transfer. 2015;**83**:604–612. DOI: 10.1016/j.ijheatmasstransfer.2014.12.044

[27] Park Y-G, Jacobi A M. Air-side heat transfer and friction correlations for flat-tube louver-fin heat exchangers. Journal of Heat Transfer. 2009;**131**(6):021801–1-11. DOI: 10.1115/1.3000609

[28] Webb R L, Jung S H. Air-side performance of enhanced brazed aluminium heat exchangers. ASHRAE Transactions. 1992;**98**(2):391–401.

[29] Rugh J P, Pearson J T, Ramadhyani S. A study of a very compact heat exchanger used for passenger compartment heating in automobiles, in compact heat exchangers for power and process industries. In: ASME Symposium Series, HTD-Vol. 201; New York: ASME; 1992. p. 15–24.

[30] Sunden B, Svantesson J. Correlation of j- and f-factors for multilouvered heat transfer surfaces. In: 3rd UK National Heat Transfer Conference; London: South Kensington Campus; 1992. p. 805–811.

[31] Tanaka T, Itoh M, Kudoh M, Tomita A. Improvement of compact heat exchangers with inclined louvered fins. Bulletin of JSME. 1984;**27**(224):219–226. DOI: 10.1299/jsme1958.27.219

[32] Jacobi A M, Park Y-G, Zhong Y, Michna G, Xia Y. High performance heat exchangers for air-conditioning and refrigeration applications (non-circular tubes). University of Illinois, Phase II-Final Report. 2005;ARTI-21CR/611–20021

[33] Dogan B, Altun Ö, Ugurlubilek N, Erbay L B. An experimental comparison of two multi-louvered fin heat exchangers with different numbers of fin rows. Applied Thermal Engineering. 2015;**91**:270–278. DOI: 10.1016/j.applthermaleng.2015.07.059

Chapter 5

Condensation Heat Transfer on Geometrically Enhanced Horizontal Tube: A Review

Hafiz Muhammad Ali

Additional information is available at the end of the chapter

http://dx.doi.org/10.5772/65896

Abstract

In this chapter, an attempt has been made to present the recent state of knowledge of free-convection condensation heat transfer on geometrically enhanced tubes. This survey is divided into three sections. The first section concentrates on research on condensate flooding or retention. The second and the third sections cover the experimental and the theoretical work on geometrically enhanced tubes, respectively.

Keywords: phase change, heat transfer, condensation, horizontal tube, geometrically enhanced tubes, enhancement, integral fins, pin fins, model

1. Introduction

The phenomenon of condensation heat transfer has been researched for over a century now. In the beginning, the primary focus to increase heat transfer was kept limited to the increase in surface area. Later, it was revealed that surface-tension forces play a vital role in thinning the condensate layer which in turn increases heat transfer. The mechanism of condensation heat transfer on two-dimensional integral-fin tubes is now well understood. Researchers have successfully identified the optimum geometries, fin shapes, dimensions and materials for integral-fin tubes for a wide range of condensing fluids. A number of theoretical models, for instance Briggs and Rose [1], Ali and Briggs [2], have successfully combined the effect of surface tension and gravity to explain condensation heat transfer on integral-fin and pin-fin tubes. However, relatively fewer investigations have been carried out for condensation on three-dimensionally enhanced tubes.

This chapter presents a state-of-the-art review of condensation heat transfer on single-horizontal geometrically enhanced tubes. The problem of condensate retention on geometrically

INTECH
open science | open minds

© 2017 The Author(s). Licensee InTech. This chapter is distributed under the terms of the Creative Commons Attribution License (http://creativecommons.org/licenses/by/3.0), which permits unrestricted use, distribution, and reproduction in any medium, provided the original work is properly cited.

enhanced tubes has been reviewed in detail followed by the experimental and theoretical investigations of condensation heat transfer on integral-finned and pin-finned tubes.

2. Free-convection condensation on horizontal smooth tubes

The first investigator to propose a theoretical model of condensation heat transfer on vertical plates and horizontal tubes was Nusselt [3]. By considering laminar flow and constant properties for the condensate film, uniform temperature on the vapour side (no temperature gradient in the vapour), and neglecting inertia, convection in the condensate film (i.e. heat transfer across the condensate film occurs only by conduction) and shear stress at the condensate surface, the following results were obtained:

For a vertical plate:

$$Nu_P = 0.943\left[\frac{\rho(\rho-\rho_v)gh_{fg}L^3}{\mu k\Delta T}\right]^{1/4} \tag{1}$$

For a horizontal tube:

$$Nu_T = 0.728\left[\frac{\rho(\rho-\rho_v)gh_{fg}d^3}{\mu k\Delta T}\right]^{1/4} \tag{2}$$

Many theoretical investigations have since been carried out including factors neglected by Nusselt [3] such as convection in condensate, shear stress and inertia (for instance, Sparrow and Gregg [4], Koh et al. [5] and Chen [6, 7]). The inclusion of these parameters made little practical difference to the results of Nusselt [3]. Rose [8] reports a comprehensive literature review of theoretical studies of laminar-film condensation on smooth tubes.

3. Free-convection condensation on horizontal-enhanced tubes

3.1. Condensate retention or flooding

It is well understood that heat-transfer rate is strongly influenced by the available area. For that reason, a long time ago smooth tubes were replaced by horizontal integral-fin tubes. No doubt, the addition of the fins provides an increase in area that ultimately leads to an enhancement in heat transfer, but a significant amount of condensate is retained on the tube due to capillary forces. This phenomenon of trapped liquid between fins is known as 'condensate retention or flooding' and is illustrated in **Figure 1**. This condensate offers a great resistance to heat transfer. A flooding angle, $\varnothing_f$, has been defined to indicate the point from the top of the tube, where the condensate flooding completely fills the inter-fin spacing up to the tip of the fin. This problem of condensate retention was first experimentally investigated by Katz et al. [9].

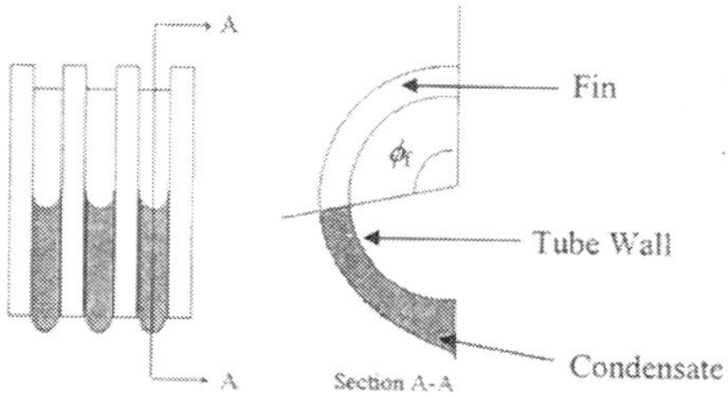

Figure 1. Condensate flooding on a horizontal integral-fin tube showing retention angle $\varnothing_f$.

Rudy and Webb [10, 11] reported experimental investigations of condensate retention on three integral-fin tubes and a spine-fin tube with fin densities in a range of 748–1378 fins per meter using n-pentane, R-11 and water under static (without condensation) and dynamic (with condensation) conditions. For all the fluids, they found an increase in condensate retention with increasing fin density and surface-tension-to-density ratio. They also found no significant differences in condensate retention under static and dynamic conditions.

Honda et al. [12] presented a comprehensive experimental and theoretical analysis of condensate flooding using R-113 and methanol on three horizontal integral-fin tubes and a saw-toothed tube with and without 'drainage strips'. A significant decrease in condensate retention was reported when the same tubes were used with drainage strips. One of the integral-fin tubes was tested under both static and dynamic conditions and no significant change in condensate retention was observed which was in line with the findings of Rudy and Webb [10].

Honda et al. [12] made the following assumptions for their theoretical analysis of the static meniscus between trapezoidal fins:

1. The meniscus is just in contact with the fin tip.
2. The radius of curvature of the condensate interface is much smaller in the longitudinal direction than in the circumferential direction.
3. The fin height (h) and the fin-tip spacing (b) are sufficiently smaller than the fin-tip radius (R_o).
4. The radius of curvature at the tube bottom is infinite in the longitudinal direction.

Using the above assumptions, the following expression was produced for retention angle, $\varnothing_f$, measured from the top of the tube:

$$\varnothing_f = \cos^{-1}\left[\left(\frac{2\sigma\cos\theta}{\rho g b R_o}\right) - 1\right] \tag{3}$$

where

$$h > (s/2)\cos\theta \tag{4}$$

Honda et al. [12] compared their own experimental data, experimental data of Katz et al. [9] and experimental data of Rudy and Webb [10] using Eq. (3). Good agreement was found between experiment and theory. Later, Rudy and Webb [11] and Owen et al. [13] obtained the same Eq. (3) of condensate-retention angle for integral-fin tubes.

Yau et al. [14] reported experimental data for condensate-retention angle using fluids steam, ethylene glycol and R-113. Thirteen tubes with rectangular integral fins were tested with a fin height of 1 mm, a thickness of 0.5 mm and a variable-fin spacing. For tubes with $h > s/2$, the measured retention angles showed good agreement using Eq. (3) as shown in **Figure 2**. Two tubes with fin spacing of 1.5 and 2 mm were also tested using copper-drainage strips; a significant increase in the retention angle was noted. The following empirical expression was determined for the retention angle of tubes using drainage strips:

$$\varnothing_f = \cos^{-1}\left(\frac{0.83\sigma}{\rho g b R_o}-1\right) \tag{5}$$

Figure 2 also shows a good agreement of experimental data using Eq. (5). A provisional equation for trapezoidal integral-fin tubes using drainage strips was also suggested as

$$\varnothing_f = \cos^{-1}\left(\frac{0.83\sigma\cos\theta}{\rho g b R_o}-1\right) \tag{6}$$

Masuda and Rose [15] comprehensively analyzed the configuration of the liquid film retained by surface-tension forces on horizontal low integral-fin tubes ($h \ll R_o$). This study revealed that

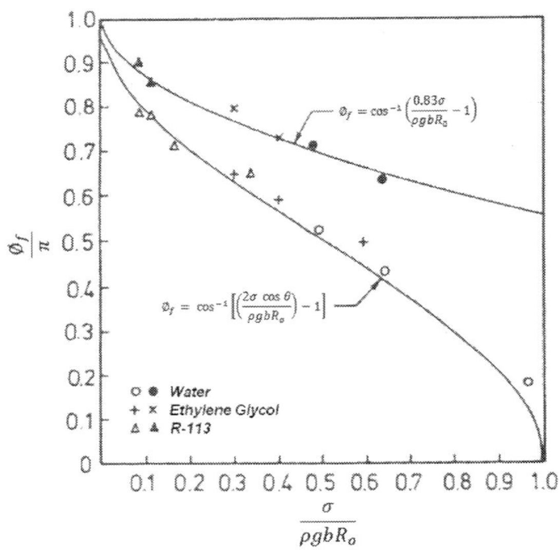

Figure 2. Liquid retention results with and without drainage strips (after Yau et al. [14]).

http://dx.doi.org/10.5772/65896

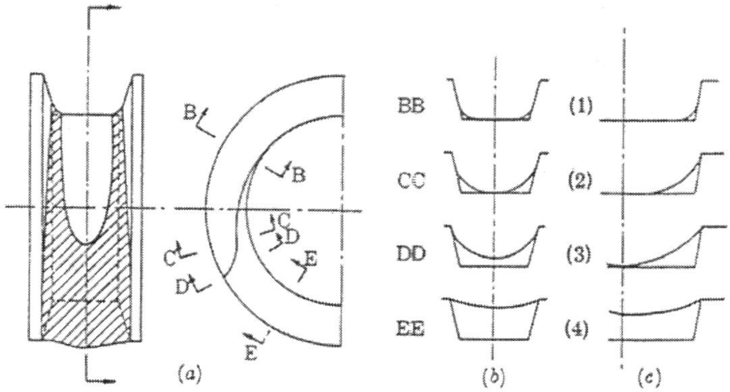

Figure 3. Configuration of retained liquid or condensate around a horizontal integral-fin tube. (a) Condensate retention on integral-fin tube. (b) Configuration of liquid around the narrow-spaced integral-fin tube. (c) Configuration of liquid around the wide-spaced integral-fin tube (after Masuda and Rose [15]).

the liquid is not only retained on the lower part of the tube (below the retention angle) but also on the upper part of the tube surface in the form of 'wedges' between fin flanks and tube surface in the inter-fin space. This phenomenon is illustrated in **Figure 3**. **Figure 3b** and **c** describes the configuration of liquid around the tube for narrow-spaced ($h > (s/2)\cos\theta$) and wide-spaced ($h < (s/2)\cos\theta$) integral-fin tubes, respectively. Four 'flooding' conditions were identified for trapezoidal fins and the positions around the tube at which these occur were determined:

For narrow-spaced fins ($h > (s/2)\cos\theta$), where the inter-fin space is just filled by the meniscus but the fin flanks are not wholly wetted (**Figure 3b.2**). Retention angle, $\varnothing_f$, for this case is given as

$$\cos\varnothing_f = \left(\frac{2\sigma}{\rho g s R_r}\right)\left[\frac{\cos\theta}{1+\sin\theta}\right]-\left(\frac{R_o}{R_r}\right) \tag{7}$$

For narrow-spaced fins ($h > (s/2)\cos\theta$), where the whole flank is just wetted (contact angle at the fin tip is zero) and for which the liquid film at the centre of the inter-fin space has finite thickness (**Figure 3b.3**). This is the condition for which Honda et al. [12] derived flooding angle Eq. (3). Retention angle, $\varnothing_f$, for this case is given as

$$\cos\varnothing_f = \left(\frac{2\sigma}{\rho g b R_o}\right)\cos\theta-1 \tag{8}$$

For wide-spaced fins ($h < (s/2)\cos\theta$), where the fin flanks are wholly wetted (contact angle at the fin tip is zero) before the inter-fin space is flooded (**Figure 3c.2**). Retention angle, $\varnothing_f$, for this case is given as

$$\cos \varnothing_f = \left(\frac{\sigma}{\rho g h R_r}\right)(1-\sin\theta)-\left(\frac{R_o}{R_r}\right) \tag{9}$$

For wide-spaced fins ($h < (s/2)\cos\theta$), where the whole of the inter-fin space is just flooded and the contact angle at the fin tip is no longer zero (**Figure 3c.3**). Retention angle, $\varnothing_f$, for this case is given as

$$\cos \varnothing_f = \left(\frac{8\sigma h}{\rho g(b^2+4h^2)R_o}\right)-1 \tag{10}$$

Masuda and Rose [15] also defined an 'active area enhancement ratio' for low-rectangular integral-fin tubes (when $h > s/2$) as 'The unblanked area of a finned tube (i.e. area of fin tips plus area of unblanked part of fin flanks plus area of unblanked part of inter-fin tube surface) divided by the area of a smooth tube with radius equal to the fin root radius'. The following equation was derived for active area-enhancement ratio, ξ,

$$\xi = \frac{R_r b\varnothing_f(1-f_s) + (R_o{}^2-R_r{}^2)\varnothing_f(1-f_f) + \pi R_o t}{\pi R_r(b+t)} \tag{11}$$

where f_s and f_f are the blanked proportions of the inter-fin space and fin flanks for the unflooded part of rectangular-finned tube, respectively, and are given by the following expressions:

$$f_s = \left(\frac{2\sigma}{\rho g b R_r}\right)\left\{\frac{\tan(\varnothing_f/2)}{\varnothing_f}\right\} \tag{12}$$

and

$$f_f = \left(\frac{\sigma}{\rho g h R_r}\right)\left\{\frac{\tan(\varnothing_f/2)}{\varnothing_f}\right\} \tag{13}$$

Rose [16] extended the work and proposed expressions for f_s and f_f for trapezoidal-finned tubes as

$$f_s = \left\{\frac{1-\tan(\theta/2)}{1+\tan(\theta/2)}\right\}\left(\frac{2\sigma}{\rho g b R_r}\right)\left\{\frac{\tan(\varnothing_f/2)}{\varnothing_f}\right\} \tag{14}$$

and

$$f_f = \left\{\frac{1-\tan(\theta/2)}{1+\tan(\theta/2)}\right\}\left(\frac{\sigma}{\rho g h R_r}\right)\left\{\frac{\tan(\varnothing_f/2)}{\varnothing_f}\right\} \tag{15}$$

It was suggested by Masuda and Rose [15] that manufacturing integral-fin tubes with filleted fin roots would replace the retained wedges of condensate with high-conductivity metal, and hence increase the active area-enhancement ratio resulting in more heat transfer. Wen et al. [17] experimentally investigated the effect of fillet fin roots on heat-transfer enhancement using steam, ethylene glycol and R-113 as condensing fluids on four integral-fin tubes. A significant enhancement was found for tubes with filleted roots over tubes without filleted roots.

Briggs [18] obtained static liquid-retention measurements on 12 three-dimensional pin-fin tubes and three integral-fin tubes. R-113, ethylene glycol and water were used as test fluids. Static retention measurements were obtained by using two methods: first by taking photographs and second by counting pins. A comparison of both methods of measuring retention angles is shown in **Figure 4**; it can be seen for water and ethylene glycol, and both methods give results within 15%, but for R-113, pin-counting method gives higher-retention angles compared to the photographic method. Finally, retention angles for water and ethylene glycol were taken as the average of both the methods, but for R-113, pin-counting method was used as it was deemed more accurate than photographic method. Liquid retention on three-dimensional pin-fin tubes was found to be lower than the equivalent integral-fin tubes (i.e. with the same longitudinal- and radial-fin dimensions). The controlling parameters appeared to be longitudinal and circumferential pin spacing. A tube with 1-mm circumferential spacing was found to be optimum for flooding angle. Pin height and longitudinal and circumferential pin thickness had little influence on retention.

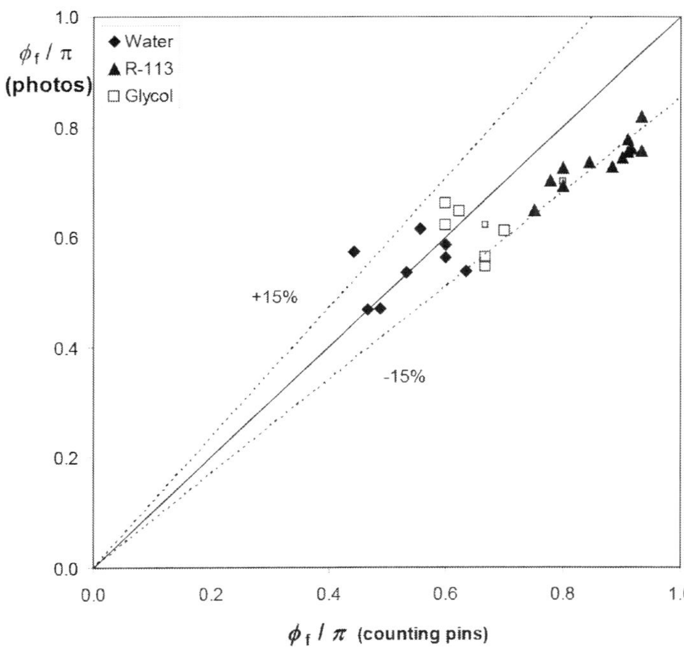

Figure 4. Comparison of pin-counting method with photographic method by Briggs [18].

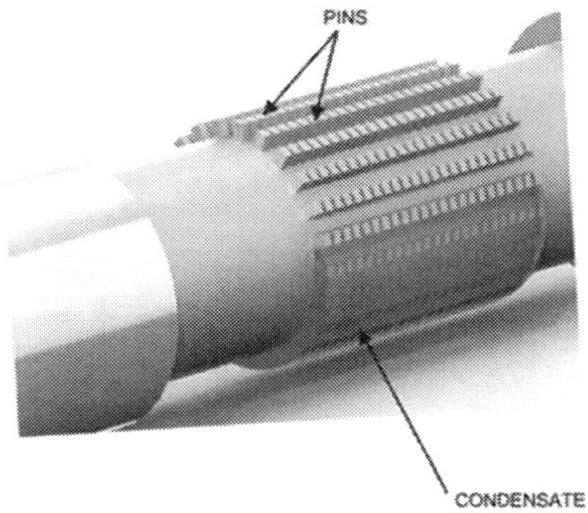

Figure 5. Sketch of a three-dimensional pin-fin tube.

Comprehensive experimental data for condensate retention (for free convection) were reported on 15 pin-fin tubes (**Figure 5**) by Ali and Briggs [19]. Static method to create condensate was adopted to carry out experimentation. Pin-counting and photographic methods were used to analyse condensate and a comparison of both methods was found to be within ±5%. All pin-fin tubes were found to be less flooded than the equivalent integral-fin tubes. A semi-empirical model was also reported for condensate-retention angle on pin-fin tubes as follows:

$$\varnothing_f = \cos^{-1}\left[\left(1-\left(0.4919-1.306\left(\sigma/_{\rho R^2 g}\right)\right)\frac{s_c}{t_c}\right)\left(\frac{2\sigma}{\rho g s R_o}\right)-1\right] for\ s < 2h \tag{16}$$

Ali and co-workers [20–22] have reported in detail studies on condensate retention as a function of vapour velocity on horizontal integral-fin and pin-fin tubes. Recently, Ali et al. [23] reported the effect of condensate flow rate on retention angle for horizontal integral-fin tubes.

3.2. Experimental studies into condensation heat transfer on enhanced tubes

Table 1 summarizes the key facts and figures of experimental investigations carried out on enhanced tubes which are described in detail in the following sections.

3.2.1. Tubes with two-dimensional fins

Honda et al. [12] presented heat-transfer measurements for the condensation of R-113 and methanol on three integral-fin tubes and a three-dimensional saw-toothed tube. The vapour-side, heat-transfer coefficient was found by direct measurements (12–16 thermocouples were placed in each tube wall). The saw-toothed tube gave the best heat-transfer enhancements (defined as heat-transfer coefficient for saw-toothed tube based on fin-tip diameter divided by the heat-transfer coefficient for a smooth tube at the same vapour-side, temperature difference) for both fluids which was 9.0 and 6.1 for R-113 and methanol, respectively.

http://dx.doi.org/10.5772/65896

Investigation	Number of tubes tested	Type of tube	Heat-transfer coefficient calculation method	Fluids tested	Maximum reported heat-transfer enhancement ratio
Honda et al. [12]	4	3 trapezoidal integral fin 1 saw-toothed	Direct measurements	R-113 Methanol	9.0 6.1
Yau et al. [24]	13	Rectangular integral fin	Predetermined coolant-side correlation	Steam	3.6
Masuda and Rose [25]	14	Rectangular integral fin	Predetermined coolant-side correlation	R-113	7.3
Masuda and Rose [26]	14	Rectangular integral fin	Predetermined coolant-side correlation	Ethylene glycol	4.4
Wanniarachchi et al. [27, 28]	24	Rectangular integral fin	Predetermined coolant-side correlation	Steam	5.2
Marto et al. [30]	24	Rectangular integral fin	Modified Wilson plot	R-113	7.0
Sukathme et al. [34]	12	9 trapezoidal integral fins 3 Pin fins	Direct measurements	R-11	10.3 (integral fin) 12.3 (pin fin)
Briggs et al. [31]	6	Rectangular integral fin	Direct measurements	R-113 ethylene glycol steam	6.8 4.8 3.0
Briggs et al. [35]	17 (commercial tubes)	7 two-dimensional 10 three-dimensional	Predetermined coolant-side correlation	R-113	8.2
Kumar et al. [37]	2	1 integral fin 1 pin fin	Modified Wilson plot and direct measurements	Steam	2.5 (integral fin) 3.6 (pin fin)
Kumar et al. [39]	8	2 integral fins 2 pin fins 4 partial integral fins	Direct measurements	Steam R-134a	2.9 6.5
Briggs [41]	6	Pin-fin tubes	Predetermined coolant-side correlation	R-113 steam	9.9 2.9
Park et al. [33]	4	Integral fin	Direct measurements	R-123	5.8
Baiser and Briggs [42]	5	Pin fin	Predetermined coolant-side correlation	Steam	4.1

Table 1. Summary of experimental literature review.

Yau et al. [24] reported an experimental study of dependence of heat transfer on fin spacing for the condensation of steam on horizontal integral-fin tubes. Thirteen tubes with rectangular fins having a thickness of 0.5mm and a height of 1.6 mm were tested by systematically varying fin spacing from 0.5 to 20mm. All tubes were having a fin-root diameter of 12.7mm. A plain tube with an outer diameter equal to the fin-root diameter was also tested for comparison. All tests were performed at near-atmospheric pressure with vapour flowing vertically downward with velocities between 0.5 and 1.1m/s. The vapour-side, heat-transfer coefficients were found by subtracting the predetermined coolant side and wall resistance from the overall thermal resistance. The observed heat-transfer enhancement for integral-fin tubes significantly exceeded the increase in active area. The maximum vapour-side, heat-transfer enhancement was found to be around 3.6 for the tube with a fin spacing of 1.5mm. Integral-fin tubes with a spacing of 0.5 and 1.0mm were found to be almost completely flooded by condensate.

Yau et al. [14] also used solid drainage strips with two integral-fin tubes of fin spacing of 1.5 and 2.0mm and found that for steam the drainage strip significantly reduced the condensate flooding. The drainage strips were made of copper having a thickness of 0.5mm and a height of 8mm. The tubes with strips provided about 25–30% additional heat-transfer enhancement compared to the same integral-fin tubes without strips.

Masuda and Rose [25, 26] reported experimental data for the condensation of R-113 and ethylene glycol on integral-fin tubes. The effect of fin spacing was investigated on the same set of tubes as used by Yau et al. [14, 24] with the inclusion of a new integral-fin tube with a fin spacing of 0.25mm. Predetermined coolant-side correlation and a modified Wilson plot method were used to evaluate the vapour-side, heat-transfer coefficients. For both condensing fluids vapour-side, heat-transfer enhancement was found to be about two times higher than the corresponding active area. Tubes with a spacing of 0.5 and 1.0mm showed best heat-transfer enhancement of 7.3 for R-113 and 4.4 for ethylene glycol, respectively.

Masuda and Rose [15] summarized the above experimental investigations by plotting the dependence of vapour-side, heat-transfer enhancement against fin spacing. For steam, ethylene glycol and R-113, tubes with a fin spacing of 1.5, 1 and 0.5mm, respectively, gave the best heat-transfer enhancement. They also plotted a graph of active-area enhancement against fin spacing. For steam, ethylene glycol and R-113, integral-fin tubes with a fin spacing of 1.5, 1 and 0.5mm gave the best active-area enhancement, respectively. Thus, heat-transfer enhancement is a maximum for fin spacing that maximize the active area.

Wanniarachchi et al. [27, 28] reported vapour-side, heat-transfer measurements for the condensation of steam at atmospheric and low (11.3kPa) pressure on 24 horizontal rectangular cross-section integral-fin tubes made of copper. Fin spacing (0.5, 1.0, 1.5, 2.0, 4.0mm), fin thickness (0.5, 0.75, 1.0 and 1.5mm) and fin height (0.5, 1.0, 1.5 and 2.0mm) were changed systematically to find the best geometry for heat transfer. Vapour-side, heat-transfer coefficients were obtained using a predetermined coolant-side correlation and also by a modified Wilson plot method. Enhancement ratio was found to be strongly dependent on fin spacing and an optimum value was reported between 1.5 and 2.0mm for all tubes. Fin thickness showed a weak effect on enhancement ratio with an optimum range between 0.75 and 1.0mm.

Enhancement ratio was found to increase with increasing fin height but at a lower rate than the area increase.

Marto et al. [29] presented an experimental study to identify the optimum fin shape to maximize heat transfer. Four integral-fin tubes with rectangular, triangular, trapezoidal and parabolic fin shapes were tested using steam as the condensing fluid. All tubes had a same fin height and fin-root spacing and thickness. Tests were carried out at near-atmospheric and below-atmospheric pressures. A tube with a roughly parabolic fin shape outperformed the tubes with rectangular, triangular and trapezoidal fin shapes at both pressures.

Marto et al. [30] reported experimental data condensing R-113 on 24 integral-fin tubes and a commercially available tube. Fin spacing was varied systematically in a range of 0.25–4 mm for different sets of fin thicknesses. All tests were performed at a little above atmospheric pressure with a downward-flowing vapour velocity of 0.4 m/s. Vapour-side, heat-transfer coefficients were obtained using the modified Wilson plot method with a measured uncertainty in the range of ±7%. The tube with a fin spacing of 0.25 mm and a thickness of 0.5 mm gave the best heat-transfer enhancement of 7 for a corresponding area enhancement of 3.9. For all tubes tested, heat-transfer enhancements were found to be considerably higher than the corresponding increase in active areas. The best fin spacing was obtained to be in between 0.2 and 0.5 mm, depending upon the corresponding fin thickness and height. Heat-transfer coefficient was also found to increase with increase in fin height, but the rate of increase in coefficient of heat transfer was found to decrease with the increase in height.

Briggs et al. [31] reported experimental data for the condensation of steam, ethylene glycol and R-113 on two sets of integral-fin tubes. The smaller tubes had a fin-root diameter of 12.7 mm, fin thickness 0.5 mm and fin height 1.6 mm, whereas the larger tubes had a fin-root diameter of 19.1 mm and fin thickness and height of 1.0 mm. For both types, three fin spacings of 0.5, 1.0 and 1.5 mm were tested. The outside tube-wall temperature was measured directly by four embedded thermocouples. For all the smaller tubes, tests were conducted at a little above atmospheric pressure. For larger tubes, tests were performed at a little above atmospheric pressure for steam and R-113 and also at lower pressures of 3 and 14 kPa for ethylene glycol and steam, respectively. For both larger and smaller diameters, the best-performing integral-fin tubes were found with fin spacings of 1.5, 1.0 and 0.5 mm for steam, ethylene glycol and R-113, respectively. They compared their own experimental data with the indirectly obtained experimental data of earlier investigators [24, 27, 28, 30] and a satisfactory agreement was found.

Briggs et al. [32] reported systematic experimental data for the condensation of steam and R-113 on rectangular integral-fin tubes made of copper, brass and bronze, with fin spacing and fin-root diameter of 1.0 and 12.7 mm, respectively; fin heights and thicknesses varied in the range of 0.5–1.6 mm and 0.25–0.75 mm, respectively. For R-113, the heat-transfer enhancement was weakly dependent on fin thermal conductivity but more strongly dependent on fin height and thickness, whereas for steam, the effect of thermal conductivity on heat-transfer enhancement was much stronger for larger fin heights, but the effect of fin height and thickness was relatively small.

Park et al. [33] obtained experimental data for R-123 condensing on four integral-fin tubes used in building chillers with varying fin density in a range of 10 fins per inch to 36 fins per inch. A plain tube with the same outside diameter was also tested to compare the results. The vapour-side, heat-transfer coefficients were found directly with embedded thermocouples in the tube wall. The tube with a fin density of 28 fins per inch was found to be optimum with a vapour-side, heat-transfer enhancement of 5.8.

3.2.2. Tubes with three-dimensional fins

Sukathme et al. [34] obtained experimental data for the condensation of R-11 on nine horizontal integral-fin tubes and three special pin-fin tubes made of copper and reported the effect of fin height, fin density and fin-tip angle on vapour-side, heat-transfer coefficients. All tubes were made with trapezoidal fin shapes. Vapour-side, heat-transfer coefficients were found from directly measured tube-wall temperatures, obtained by placing 15 thermocouples at 5 positions along the tube and at top, bottom and mid-plane around the tube. Fin-tip angle showed a small effect on the vapour-side, heat transfer, whereas fin density and fin height showed considerable effects on the vapour-side, heat-transfer coefficient. The best-performing integral-fin tube with a fin density of 1417 fins per meter, a fin height of 1.22 mm and a fin-tip angle of 10° gave a vapour-side, heat-transfer of 10.3 with a corresponding active-area enhancement of 7. Further, 80 longitudinal trapezoidal grooves were machined in the best-performing integral-fin tube with three different heights of 0.7, 0.9 and 1.22 mm. The authors reported a large increase in vapour-side, heat-transfer enhancements with increasing value of height. The pin-fin tube with a longitudinal groove height of 1.22 mm gave a heat-transfer enhancement of 12.3 which was about 20% more than the equivalent best-performing integral-fin tube. The authors suggested that this increase in heat-transfer enhancement could be due to the increase in the flooding angle of the pin-fin tube which was about 20% more than the corresponding integral-fin tube.

Briggs et al. [35] reported experimental data for the condensation of R-113 on 17 commercially available copper integral-fin tubes. These consisted of seven two-dimensional tubes (Gewa N and K, trapezoidal cross section) and ten three-dimensional tubes (one thermoexcel and nine petal shaped). It was found that the best two-dimensional tube (K-50) and best three-dimensional tube (P8) gave similar vapour-side, heat-transfer enhancement of 8.2.

Cheng et al. [36] obtained condensing data for R-22 on six commercially available tubes. Two tubes had low integral fin, whereas four were three-dimensionally enhanced. One set of tubes consisting of an integral-fin tube, an externally enhanced tube and an externally plus internally enhanced tube has a fin density of 26 fins per inch, fin pitch of 0.97 mm and a height of 1.3 mm, whereas the other set of tubes consisting of one integral-fin tube, one externally enhanced tube and one externally plus internally enhanced tube has a fin density of 40 fins per inch, fin pitch of 0.61 mm and a fin height of 1.42 mm. Experiments were carried out at three different pressures of 1.3, 1.5 and 1.6 MPa. A Wilson plot method was used to obtain vapour-side, heat-transfer coefficients. The three-dimensional externally plus internally enhanced tubes showed the highest heat-transfer coefficients compared to rest of the tubes. The heat-transfer coefficients were found to decrease with increasing value of pressure. It was also found that vapour-side, heat-transfer coefficients decreased more sharply for three-dimensionally enhanced tubes as a function of increasing temperature difference compared to integral-fin tubes.

Kumar et al. [37] reported experimental data for the condensation of steam on a plain tube with an outside diameter of 22 mm and an integral-fin tube (with an outside diameter of 25 mm, fin height of 1.1 mm, fin thickness of 1.1 mm and fin spacing of 1.5 mm). A three-dimensional pin-fin tube was also tested with similar radial and longitudinal dimensions as of integral-fin tube but with 40 axial grooves around the circumference producing a circumferential pin spacing of 0.9 mm. The condensing-side heat-transfer coefficients were found using a modified Wilson plot method and also by direct measurement of wall temperatures; good agreement was found between the two methods. Vapour-side, heat-transfer enhancements of 2.5 and 3.6 were found for the integral-fin tube and pin-fin tube, respectively. The superior performance of the pin-fin tube was thought to be primarily due to the thinning of the condensate film by the surface-tension pull in two directions in the unflooded area as also proposed by Sukhatme et al. [34] condensing R-11 and also due to the improved condensate drainage at the bottom of the tube. Authors reported the improved condensate drainage at the bottom part of the pin-fin tube compared to the condensate drainage for the integral-fin tube.

Jung et al. [38] reported vapour-side, heat-transfer enhancements for an integral-fin tube with fin density of 26 fins per inch and a three-dimensional turbo-C tube with a fin density of 42 fins per inch condensing two low-pressure (R-11 and R-123) and two medium-pressure (R-12 and R-134a) refrigerants. A plain tube was also tested for comparison. Vapour-side, heat-transfer coefficients were obtained directly by measuring the tube-wall temperature with embedded thermocouples. For low-pressure refrigerants, heat-transfer coefficients for R-123 of about 8–19% lower than those of R-11 were found for all tubes tested. For medium-pressure refrigerants, heat-transfer coefficients for R-134a were about 0–32% higher than those for R-12. The vapour-side, heat-transfer enhancements for turbo-C and integral-fin tubes based upon the plain tube area were roughly reported up to 8.0 and 5.5, respectively.

Kumar et al. [39, 40] presented experimental data for the condensation of steam and R-134a. Five tubes consisting of one plain, one integral-fin, one pin-fin and two partial integral-fin tubes (i.e. one with pin fins on the upper half and one with pin fins on the lower half) were tested for each fluid. For steam, all enhanced tubes had rectangular fins and a fin density of 390 fins per meter, whereas for R-134a, all enhanced tubes had trapezoidal fins and a fin density of 1560 fins per meter. Pin-fin tubes were made by machining longitudinal grooves into integral-fin tubes. Pin-fin tubes gave the best vapour-side, heat-transfer enhancements of 2.9 for steam (30% more than equivalent integral-fin tube tested) and 6.5 for R-134a (24% more than equivalent integral-fin tube tested). Pin fins were reported to be more effective at lower half of the tube than the upper half of the tube, that is, for steam, a heat-transfer enhancement of 2.4 (with pin fin on the upper half) and 2.7 (with pin fins on the lower half), whereas for R-134a, a heat-transfer enhancement of 5.7 (with pin fins on the upper half) and 6.3 (with pin fins on the lower half) was reported. Tubes with pin fins on the lower half outperformed the equivalent integral-fin tubes by up to 20% for steam and 11% for R-134a. For R-134a, pin fins on the upper half of the tube did not contribute in the heat-transfer enhancement but showed 5% improvement for steam compared to integral-fin tube.

Briggs [41] reported experimental data for the condensation of R-113 and steam on six three-dimensional pin-fin tubes. These tubes were made by machining rectangular longitudinal grooves into integral-fin tubes. A plain tube with the same outside diameter as the pin-fin tube-root diameter was also tested for comparison purposes. The vapour-side, heat-transfer coefficient was obtained by subtracting the coolant and wall resistances from the measured overall resistance. For R-113, the best-performing tube had circumferential pin thickness and spacing of 0.5mm, pin height of 1.6mm and a longitudinal spacing and thickness of 0.5mm. For steam, the best-performing tube had circumferential pin thickness and spacing of 0.5 and 1.0mm, respectively, and longitudinal thickness of 0.5mm and spacing of 1.1mm. Tubes with larger fin heights produced higher heat transfer when all other geometric variables remained the same. For R-113, the best-performing tube gave a vapour-side enhancement of 9.9 compared to the plain tube which was about 40% higher than the equivalent integral-fin tube with the same fin height, longitudinal thickness and spacing. For steam, the best-performing tube gave a heat-transfer enhancement of 2.9 compared to the plain tube which was about 25% higher than the equivalent integral-fin tube. For R-113, a near-linear increase in heat-transfer enhancement with active-area enhancement was reported. The heat-transfer enhancement was approximately twice the active-area enhancement. For steam, heat-transfer enhancement was virtually independent of active-area enhancement. The author also reported that static condensate flooding on pin-fin tubes was significantly less than the equivalent integral-fin tubes.

Baiser and Briggs [42] reported experimental data for the condensation of steam at atmospheric pressure and low velocity on five three-dimensional copper pin-fin tubes. These were the same tubes used in the investigations of Briggs [18]. All of the tubes had a pin-fin root diameter of 12.7 mm. Only circumferential thickness and spacing were varied. Vapour-side, heat-transfer coefficients were found by subtracting the coolant and wall resistances from the measured overall thermal resistance. All pin-fin tubes gave higher vapour-side, heat-transfer coefficients compared to the equivalent integral-fin tube. The best heat-transfer enhancement was found to be 4.1 which was thought to be on par with the best-reported heat-transfer enhancement on an optimum integral-fin tube by Wanniarachchi et al. [28]. It was noted that despite less active area of the pin-fin tubes compared to the equivalent integral-fin tube, pin-fin tubes outperformed the integral-fin tube. It was suggested due to the fact that in the case of pin-fin tubes, many small effective surfaces replaced few large surfaces of integral-fin tubes and these smaller surfaces are far more effective for heat transfer since in gravity-drained flows, they result in shorter thinner boundary layers, while for surface-tension-driven flows, these small surfaces produce many more sharp changes in surface curvature, which result in surface-tension-induced pressure gradients which thin the condensate film. An optimum circumferential spacing of 1mm was also identified which maximized the heat-transfer rate.

Ali and Briggs [43–46] have reported a comprehensive data for the condensation of R-113 and ethylene glycol on various pin-fin tubes. Their work has shown superior heat-transfer performance of pin-fin tubes (up to 25%) over the equivalent integral-fin tubes (i.e. with the same fin height, root diameter and longitudinal pin thickness and spacing).

Another useful method to enhance heat transfer on horizontal tubes is by wrapping the wire on the smooth tube; recently, studies are reported by Ali and Qasim [47, 48].

3.3. Theoretical studies into condensation heat transfer on enhanced tubes

3.3.1. Tubes with two-dimensional fins

Beatty and Katz [49] were the first to propose a model for condensation heat transfer on integral-fin tubes. The model assumed the following points:

1. Gravity drains the condensate from the vertical fins and from the tube in the inter-fin spacing.
2. Surface-tension effects were entirely ignored, that is, the model did not account for capillary retention on the lower part of the tube or enhanced drainage due to surface tension on the upper part of the tube.
3. The model ignored condensation on the fin tips.

Condensation on the vertical fin flanks was modelled by applying the Nusselt [3] equation for vertical plates and condensation in the inter-fin spacing was modelled by applying the Nusselt [3] equation for horizontal tubes. The mean vapour-side, heat-transfer coefficient for the integral-fin tube was calculated as the area-weighted average of the heat-transfer coefficient on finned surfaces and on base tube between inter-fin spacing. The following expression was suggested for the vapour-side, heat-transfer coefficient:

$$\alpha = c\left[\frac{k^3\rho^2 g h_{fg}}{\mu \Delta T}\right]^{1/4}\left[\frac{A_r}{A_d}d^{-1/4} + 1.3\frac{A_f}{A_d}L_f^{-1/4}\right] \tag{17}$$

where c is an empirical constant and when was taken as 0.689 by Beatty and Katz, their experimental data for six low-surface-tension fluids (methyl chloride, SO_2, R-22, n-pentane, propane and n-butane) condensing on several integral-fin tubes (with fin densities from 433 to 633 fpm) were predicted within ±11%. L_f is the effective fin height (average vertical fin height over the diameter d_o), and Beatty and Katz took it as

$$L_f = \frac{\pi}{4}\left[\frac{d_o^2 - d^2}{d_o}\right] \tag{18}$$

Rose [16] pointed out that if the condensate drained from the fin flanks to the inter-fin space and proceeded to drain around the inter-fin tube surface to the bottom of the tube, then a more appropriate value of the effective fin height would be half of the Beatty and Katz [49] value giving

$$L_f = \frac{\pi}{8}\left[\frac{d_o^2 - d^2}{d_o}\right] \tag{19}$$

Briggs and Rose [50] compared the Beatty and Katz [49] model to the results of many of the experimental investigations on integral-fin tubes discussed above. The model showed acceptable agreement for relatively low surface-tension fluids but over-predicted the data for high surface-tension fluids such as steam and ethylene glycol. The authors explained that this was due to the neglect of surface-tension effects in the model.

Gregorig [51] discussed the effect of surface tension and pointed out its vital role in enhancing condensation heat transfer. His work addressed a vertical fluted surface; a schematic is shown in **Figure 6**. The author reported that surface-tension forces are the dominating factor in determining the heat transfer for fins with a height less than 1.5mm, as surface-tension induced pressure gradients due to the variation in the curvature of the vapour-liquid interface of the condensate on the fin. This induced pressure gradient would drain the condensate in the horizontal direction, over the arc length S_m (see **Figure 6**). The gravity then drains the accumulated condensate from the channels between the flutes. The pressure gradient in the horizontal direction is given by

$$\frac{dP}{dS} = \sigma \frac{d}{dS}\left(\frac{1}{r}\right) \tag{20}$$

where S is the distance along the vapour-liquid interface from the tip of the fin and r is the radius of curvature of the liquid-vapour interface. Gregorig [51] also gave a relation that described the shape of a convex profile which provides a constant condensate film thickness over the arc length S_m,

$$\frac{1}{r} = \frac{1.5\beta_m}{S_m}\left[1\text{-}\left(\frac{S}{S_m}\right)^2\right] \tag{21}$$

Adamek [52] defined a family of convex shapes that use surface tension to drain the film. His fin curvature was defined as

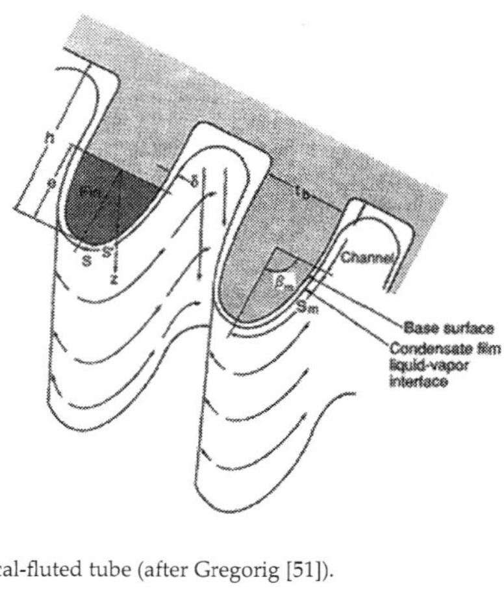

Figure 6. Fin parameters of vertical-fluted tube (after Gregorig [51]).

$$\frac{1}{r} = \frac{\beta_m}{S_m}\left(\frac{\zeta+1}{\zeta}\right)\left[1-\left(\frac{S}{S_m}\right)^{\zeta}\right] fcr \; -1 < \zeta < \infty \tag{22}$$

where each value of ζ gives a different shape of fin profile and a different aspect ratio e/t_b. The Adamek [52] profile for a value of $\zeta = 2$ is identical to the Gregorig [51] profile.

Kedzierski and Webb [53] validated the Gregorig [51] and Adamek [52] theoretical findings. Using an electrostatic discharge-machining method with a numerical-controlled machine head, they produced fin profiles for ζ = 2 and -0.5. R-11 was used as condensing fluid and experimental data agreed with the predictions to within 5%.

Rudy and Webb [54] presented a model to predict condensation heat-transfer coefficient including the surface-tension effects on fin flanks. Heat transfer through the part of the tube below the flooding angle was not considered. They totally ignored body-forces (gravity) effects on the fin flanks and assumed a constant pressure gradient due to surface tension draining the condensate from the fin flanks into the inter-fin spacing. They took the radius of the curvature of the condensate surface at the fin tip and fin root as half the fin-tip thickness and fin-root spacing, respectively. The result was the following expression for the pressure gradient on the fin flanks:

$$\frac{dP}{dx} = \frac{2\sigma}{h}\left(\frac{1}{s}+\frac{1}{t}\right) \tag{23}$$

Using the above expression to replace the body-force term in the Nusselt expression for the fin flanks, the following result was proposed for vapour-side, heat-transfer coefficient:

$$\alpha = 0.728\frac{\varnothing_f}{\pi}\left(\frac{k^3\rho h_{fg}}{\mu \Delta T}\right)^{1/4}\left[\frac{A_r}{A_d}\left(\frac{\rho g}{d}\right)^{1/4} + 1.3\frac{A_f}{A_d}\left\{\frac{2\sigma}{h^2}\left(\frac{1}{s}+\frac{1}{t}\right)\right\}^{1/4}\right] \tag{24}$$

Honda and Nozu [55] provided a prediction method for heat transfer on horizontal trapezoidal integral-fin tubes. It was pointed out by the authors that an important factor, which had been ignored in earlier theoretical models, is the non-uniformity of wall temperature, due to the large difference in heat-transfer coefficients between the unflooded and flooded regions. Their model incorporated surface tension, gravity and variable wall-temperature effects. The final expression for average heat-transfer coefficient is based on two regions: unflooded and flooded. A numerical analysis has been given just for thin film with the help of the following assumptions:

1. The wall temperature is uniform along the fin.
2. The condensate flow is laminar.
3. The condensate film thickness δ is so small that the inertia term in the momentum equation and the convection term in the energy equation can be neglected.
4. Circumferential flow on the flanks can be neglected in comparison with radial flow.

5. Fin height is substantially smaller than the tube outer radius.

The following expression was developed for the condensate film thickness along the fin:

$$\frac{\rho}{3\mu}\frac{d}{dx}\left\{\left(\rho g f_x - \sigma\frac{d}{dx}\left(\frac{1}{r}\right)\right)\delta^3\right\} = \frac{k\,\Delta\,T}{\delta h_{fg}} \tag{25}$$

f_x is the normalized component of gravity and r is the radius of curvature of condensate. It should be noted that the analysis is just given for the so-called thin-film regions. For unflooded region, this includes the fin tip, fin corner and fin flank (but not the inter-fin base), whereas for the flooded region it includes only the fin tip and fin corner.

Finally, the following expression was developed for the average Nusselt number for horizontal integral-fin tubes:

$$\frac{\mathrm{Nu}_d = \mathrm{Nu}_{du}\eta_u(1-\tilde{T}_{wu})\frac{\varnothing_f}{\pi} + \mathrm{Nu}_{df}\eta_f(1-\tilde{T}_{wf})\left(1-\frac{\varnothing_f}{\pi}\right)}{(1-\tilde{T}_{wu})\frac{\varnothing_f}{\pi} + (1-\tilde{T}_{wf})\left(1-\frac{\varnothing_f}{\pi}\right)} \tag{26}$$

$\tilde{T}_{wu}$ and $\tilde{T}_{wf}$ are the dimensionless average wall temperatures at the fin roots in the unflooded and flooded regions, respectively, and can be determined by solving the problem for circumferential wall conduction by assuming constant heat-transfer coefficients for the inner surface and for unflooded and flooded regions on the outer surface and neglecting the interaction with radial conduction. η_u and η_f are the fin efficiencies in the unflooded and flooded regions, respectively. Nu_{du} and Nu_{df} are Nusselt numbers for unflooded and flooded regions, respectively.

Honda and Nozu [55] compared their theoretical model with their own experimental data for the condensation of R-113 and methanol on three integral-fin tubes (see Honda et al. [12]) and found agreement within ±10%. The same experimental data gave agreement with Beatty and Katz [49] model within ±20%. They also compared their theoretical model with the experimental results of previous investigators including for 11 fluids and 22 tubes and found an agreement within ±20%. Briggs and Rose [50] compared the Honda and Nozu [55] model with a range of experimental data of previous investigators and reported that most of the data agreed with the model to within 25%.

Rose [16] pointed out that in most of the proposed heat-transfer models, either gravity was completely neglected when surface-tension-driven drainage was considered on the fin flanks or only the radial component was included. He also suggested the need for a simple heat-transfer model, in the form of an algebraic expression akin of Beatty and Katz [49], but including surface-tension effects. Applying dimensional analysis, the following expression for the mean condensate film thickness was proposed that accounts for both gravity and surface-tension effects:

$$\delta = \left[\frac{\mu\left(\frac{q}{h_{fg}\rho}\right)}{\frac{A(\rho-\rho_v)g}{x_g} + \frac{B\sigma}{x_\sigma^3}}\right]^{1/3} \tag{27}$$

A and B are constants and found separately for the fin tips, fin flanks and inter-fin space. x_g and x_σ are characteristic lengths for gravity and surface-tension-driven flows, respectively. These characteristic lengths are different for gravity and surface-tension-driven flows. Also, the mean heat flux, q, through the condensate assuming radial conduction is given as

$$q = \frac{k\Delta T}{\delta} \tag{28}$$

For the fin tip, where there is no retained condensate, the author took the parameters involved in Eq. (27) as $A = 0.728^4$, $x_g = d_o$, $x_\sigma = t$ and $B = B_t$ (to be found empirically).

For the unflooded part of the fin flanks, the author took the parameters in Eq. (27) as $A = 0.943^4$, $x_g = h_v$, $x_\sigma = h$, $B = B_f$ (to be found empirically). h_v is the mean vertical fin height and was approximated as

$$h_v = \frac{h\varnothing_f}{\sin\varnothing_f} \quad for\ \varnothing_f \leq \frac{\pi}{2} \tag{29}$$

$$h_v = \frac{h\varnothing_f}{2 - \sin\varnothing_f} \quad for\ \varnothing_f > \frac{\pi}{2} \tag{30}$$

where $\varnothing_f$ is the flooding angle measured from the top of the tube.

Finally, for the unflooded part of the tube inter-fin space, the author took the parameters in Eq. (27) as $A = \{\xi(\varnothing)\}^3$, $x_g = d$, $x_\sigma = s$, $B = B_s$ (to be found empirically). The function $\xi(\varnothing)$ was approximated as

$$\xi(\varnothing) = 0.874 + 0.199110^{-2}\varnothing - 0.264210^{-1}\varnothing^2 + 0.553010^{-2}\varnothing^3 - 0.136310^{-2}\varnothing^4 \tag{31}$$

From Eqs. (27) and (28) with the appropriate values of A, B, x_g, and x_σ and neglecting temperature drop in the fin, the mean surface heat flux for the fin tip, fin flank and inter-fin space is given as

$$q_{tip} = \left\{\frac{\rho h_{fg} k^3 \Delta T^3}{\mu}\left(\frac{0.728^4(\rho - \rho_v)g}{d_o} + \frac{B_t\sigma}{t^3}\right)\right\}^{1/4} \tag{32}$$

$$q_{flank} = \left\{\frac{\rho h_{fg} k^3 \Delta T^3}{\mu}\left(\frac{0.943^4(\rho - \rho_v)g}{h_v} + \frac{B_f\sigma}{h^3}\right)\right\}^{1/4} \tag{33}$$

$$q_{\text{int}} = \left\{\frac{\rho h_{fg} k^3 \Delta T^3}{\mu}\left(\frac{\{\xi(\varnothing)\}^3(\rho - \rho_v)g}{d} + \frac{B_s\sigma}{s^3}\right)\right\}^{1/4} \tag{34}$$

From Nusselt [3], the expression for the heat flux for a plain tube is

$$q_{\text{plain}} = 0.728\left\{\frac{\rho h_{fg}k^3\Delta T^3}{\mu}\left(\frac{(\rho-\rho_v)g}{d}\right)\right\}^{1/4} \tag{35}$$

Further, assuming no heat transfer to the flooded and blanked part of fin flanks and inter-fin space, an enhancement ratio for a pitch length of trapezoidal integral-fin tube over the plain tube at the same temperature difference was obtained as

$$\varepsilon_{\Delta T} = \frac{q_{\text{tip}}\pi d_o t + \frac{\varnothing_f}{\pi}\left\{\frac{q_{\text{flank}}\pi(d_o^2-d^2)(1-f_f)}{2\cos\theta} + q_{\text{int}}\pi ds(1-f_s)\right\}}{q_{\text{plain}}\pi d(s+t)} \tag{36}$$

Finally, by substituting Eqs. (32)–(35) for the mean heat flux for fin tip, fin flanks and inter-fin space into Eq. (36), the following final expression is obtained:

$$\begin{aligned}\varepsilon_{\Delta T} = {} & \frac{d_o t}{d(b+t)}\left(\frac{d}{d_o} + \frac{B_t\sigma d}{0.728^4(\rho-\rho_v)gt^3}\right)^{1/4} \\ & + \left(\frac{\varnothing_f}{\pi}\right)\left\{\frac{(1-f_f)(d_o^2-d^2)}{2d(b+t)\cos\theta}\right\}\left\{\left(\frac{0.943}{0.728}\right)^4\left(\frac{d}{h_v}\right) + \frac{B_f\sigma d}{0.728^4(\rho-\rho_v)gh^3}\right\}^{1/4} \\ & + \left(\frac{\varnothing_f}{\pi}\right)\left\{\frac{(1-f_s)sB_l}{b+t}\right\}\left\{\frac{\{\xi(\varnothing)\}^3}{0.728^4} + \frac{B_s\sigma d}{0.728^4(\rho-\rho_v)gs^3}\right\}^{1/4}\end{aligned} \tag{37}$$

In the above expression, to account for the fact that condensate drainage from the fin flanks would affect both gravity and surface-tension contributions to the heat transfer at the inter-fin tube space, a lead constant, B_l, was introduced in the last term. Moreover, the constants B_t, B_f and B_s did not differ greatly when found separately, which led to the decision to set these constants equal. Using $B_t = B_f = B_s = 0.143$ and $B_l = 2.96$, the model predicted the dependence of enhancement ratio on fin spacing, fin thickness and fin height excellently. The author pointed out that as the model neglected conduction in the fin, the validity of the model was expected to decrease with decreasing thermal conductivity of the tube material and also with increasing slenderness ratio (h/t) of the fin. Briggs and Rose [50] compared the model with a large range of experimental data reported by different investigators. The model predicted most of the data for copper tubes in a range of 20%, whereas it poorly performed for steam condensing on tubes made of bronze with lower thermal conductivity where heat-transfer enhancement was overestimated.

Briggs and Rose [1] incorporated 'fin efficiency' effects into the model of Rose [16] in an approximate way. This was done by dividing the tube into flooded and unflooded parts. For the flooded part, the fin flanks were assumed adiabatic to find the heat flux through the fin tip, $q_{\text{tip,flood}}$. For the unflooded part, the heat flux for inter-fin space, q_{int}, was found using Eq. (34). For the fin flanks and the fin tip in the unflooded part to account for the temperature variations, 'slender fin' approximation for the conduction problem was used as follows:

$$\frac{\Delta T(x)}{\Delta T} = \frac{\cosh[m(h-x)] + (\alpha_{tip}/mk_w)\sinh[m(h-x)]}{\cosh(mh) + (\alpha_{tip}/mk_w)\sinh(mh)} \tag{38}$$

where

$$m = \sqrt{\left(\frac{2\alpha_{\text{flank}}}{k_w t}\right)} \tag{39}$$

With the help of the above equations, appropriate expressions including temperature variations for flank heat flux, q_{flank}, and tip heat flux, q_{tip}, was found for the unflooded area.

Finally, the following expression was proposed to calculate vapour-side, heat-transfer enhancement ratio:

$$\varepsilon_{\Delta T} = \frac{(\pi-\varnothing_f)d_o t q_{\text{tip, flood}} + \varnothing_f\left\{d_o t q_{\text{tip}} + (1-f_f)\frac{(d_o^2-d^2)}{2}q_{\text{flank}}\right\} + \varnothing_f(1-f_s)dsq_{\text{int}}}{q_{\text{plain}}\pi d(s+t)} \tag{40}$$

In the numerator of Eq. (40), the first term shows heat-transfer rate through the flooded part of the tube, the second term shows heat-transfer rate through the unflooded part of the fin and the third term shows heat-transfer rate through the unflooded part of fin spacing. Briggs and Rose [50] compared the experimental data from different investigations to the predictions of the Briggs and Rose [1] model. The inclusion of conduction in the fins on the basis of 'slender fin theory' improved the agreement of experimental data for low thermal conductivity tubes (i.e. bronze tubes condensing steam) with the model.

3.3.2. *Tubes with three-dimensional fins*

Kumar et al. [40] pointed out that almost all the reported heat-transfer models refer to condensation on integral-fin tubes and there was no analytical model for condensation on pin-fin or spine integral-fin tubes. They proposed a generalized empirical model to predict the vapour-side, heat-transfer coefficient for integral-fin as well as pin-fin tubes. They assumed that the heat-transfer coefficient was a function of fluid properties, tube geometry and condensate mass flow rate. This resulted in an expression for the vapour-side, heat-transfer coefficient as follows:

$$\alpha = 0.024(Re)^{-0.333}[We]^{0.3}(Y)^{1.4}\left(\frac{k^3\rho^2 g}{\mu^2}\right)^{0.333} \tag{41}$$

where all the constants in Eq. (41) were found empirically using least-square method, *Re* is the condensate film Reynolds number given by

$$Re = \frac{4\dot{m}}{\mu p} \tag{42}$$

We is the Weber number, the ratio of surface tension and inertia forces in the condensate, and for pin-fin tubes was estimated as a Pythagorean sum of the Weber numbers for the two perpendicular faces of the pins as follows:

$$We = \sqrt{We_l^2 + We_c^2} \quad (43)$$

where We_l and We_c are the Webber numbers for the longitudinal and circumferential faces of the pin and are calculated from

$$We_l = \frac{4\sigma\left(\frac{1}{t} + \frac{1}{s}\right)}{h\rho g} \quad (44a)$$

$$We_c = \frac{4\sigma\left(\frac{1}{t_c} + \frac{1}{s_c}\right)}{h\rho g} \quad (44b)$$

Note that for integral-fin tubes, only longitudinal Weber number is used in Eq. (41).

Y is a function of the tube geometry and is given by

$$Y = \frac{4A_o}{dp} = \frac{4\pi}{dp}\left[\frac{d_o^2 - d^2}{2} + d_o t + d(p - t_b)\right] \quad (45)$$

Kumar et al. [40] compared their own experimental data-condensing steam on two tubes (one integral fin and one pin fin) and R-134a on five tubes (four integral fins and one pin fin) with the model and reported an agreement within 15% for most of the experimental data. Cavallini et al. [56] compared the model to the experimental data for the condensation of steam and refrigerants on integral-fin tubes reported by previous researchers and concluded that the model was not appropriate for tubes with heights of more than 1.1 mm and with fin pitches of more than 1.0 mm or less than 0.5mm for refrigerants and less than 2.0 mm for steam. Namasivayam [57] also compared the model to the experimental data of steam and R-113 on integral-fin tubes and agreed with the conclusions of Cavallini et al. [56].

Belghazi et al. [58] presented a model for a specially designed three-dimensional Gewa C+ tube containing notches around the fin. The tube circumference was divided into flooded and unflooded regions. The authors further divided fin pitch into four regions. It was assumed that for certain regions (i.e. the regions between notches and above notches) surface tension will be the draining force and for other regions (i.e. the region below notches and inter-fin tube space) gravity will be the draining force. Nusselt [3] theory was applied to find the heat-transfer coefficients for the gravity-based drainage regions. By replacing (ρg) in the Nusselt theory with the following expression, surface-tension effects were included in their model:

$$\frac{dF_\sigma}{dV} \approx \frac{\sigma}{h}\left(\frac{1}{r_b} - \frac{1}{r_t}\right) \quad (46)$$

where $\frac{dF_\sigma}{dV}$ is the volume force, r_b and r_t are the radii of curvature of the condensate film liquid-vapour interface at the fin bottom and fin tip, respectively.

The authors compared their model with their own experimental data, for R-134a condensing on a Gewa C+ tube. The model predicted most of the experimental data to within –10%. This model uses a linear pressure variation technique to account for surface-tension effects and totally ignores gravity effects on the fin notches and above, but still shows a good agreement with experimental data. This might be due to overestimation of surface-tension effects compensating for the absence of gravity.

Ali and Briggs [2] developed a simple semi-empirical correlation accounting for the combined effect of gravity and surface tension for condensation on horizontal pin-fin tubes. The model divided the heat-transfer surface into five regions, that is, two types of pin flank, two types of pin root and the pin tip (**Figure** 7). The following equation was proposed to calculate heat-transfer enhancement:

$$\begin{aligned}\varepsilon_{\Delta T} = {} & \frac{2t_c t}{0.728\pi d(s+t)}\sum_{i=1}^{n/2}\left[\left\{0.943^4 \sin\varnothing\frac{d}{t_c} + B_{tip}\frac{d}{\left(\frac{t_c t}{2(t_c+t)}\right)^3}\frac{\sigma}{\tilde{\rho}g}\right\}\right]^{1/4} \\ & + \frac{4ht}{0.728\pi d(s+t)}\sum_{i=1}^{j/2}\left[\left\{0.943^4 |\cos\varnothing|\frac{d}{h} + B_{\mathrm{flank1}}\frac{d}{\left(\frac{ht}{2(h+t)}\right)^3}\frac{\sigma}{\tilde{\rho}g}\right\}\right]^{1/4} \\ & + \frac{4ht_c}{0.728\pi d(s+t)}\sum_{i=1}^{j/2}\left[\left\{0.943^4\frac{d}{\frac{ht_c}{|\sqrt{h^2+t_c{}^2}\sin(\varnothing+\beta)}} + B_{\mathrm{flank2}}\frac{d}{\left(\frac{ht_c}{2(t_c+h)}\right)^3}\frac{\sigma}{\tilde{\rho}g}\right\}\right]^{1/4} \\ & + \frac{\varnothing_f s}{0.728\pi(s+t)}\left[\left\{\{\xi(\varnothing_f)\}^3 + B_{\mathrm{root1}}\frac{d}{\left(\frac{\varnothing_f ds}{2t_c j}\right)^3}\frac{\sigma}{\tilde{\rho}g}\right\}\right]^{1/4} \\ & + \frac{2s_c t}{0.728\pi(s+t)}\sum_{i=1}^{j/2}\left[\left\{0.943^4\sin\varnothing\frac{d}{s_c} + B_{\mathrm{root2}}\frac{d}{\left(\frac{s_c}{2}\right)^3}\frac{\sigma}{\tilde{\rho}g}\right\}\right]^{1/4}\end{aligned} \tag{47}$$

In Eq. (47), $\varnothing$, β and j can be calculated using Eqs. (48)–(50). Only two thermophysical properties are involved in the expression of enhancement ratio, that is, surface tension, σ, and condensate density, ρ.

$$\varnothing = \frac{i}{n/2}\pi \tag{48}$$

$$\beta = \tan^{-1}(t_c/h) \tag{49}$$

$$j = n\frac{\varnothing_f}{\pi} \tag{50}$$

The model gave good overall agreement to within ±20% with the experimental data, as well as correctly predicted the dependence of heat-transfer enhancement on the various geometric parameters and fluid types.

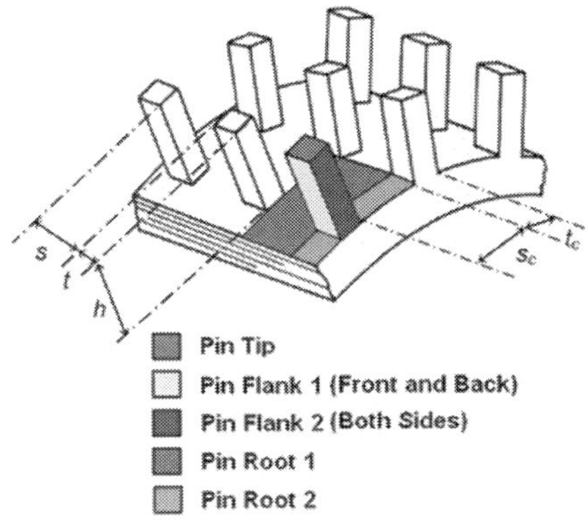

Figure 7. Schematic representation of pin-fin tube identifying five regions for heat transfer (after Ali and Briggs [2]).

4. Conclusions

Extensive experimental work has been performed on integral-fin tubes and has shown that geometry is not the only point of interest for the enhancement of heat transfer. Researchers have reported the optimum fin dimensions for a range of condensing fluids [24–26, 28, 30]. The work of Honda et al. [12] successfully predicts the condensate retention on integral-fin tubes. Reliable heat-transfer models (e.g. [1, 55]) accounting for the combined effects of surface tension and gravity on heat transfer have been developed and are readily available for design engineers.

A reasonable amount of experimental work is reported on condensation heat transfer on enhanced pin-fin tubes. Work of previous researchers has shown the superior performance of such tubes over equivalent integral-fin tubes. The extent of condensate retention and formation of many sharp surfaces enhancing surface-tension effects on pin-fin tubes are identified to be the important parameters contributing towards the heat-transfer enhancement. The model presented by Ali and Briggs [2] is available to predict heat transfer on the pin-fin tubes reasonably by accounting the effect of both gravity and surface-tension condensate drainage.

Nomenclature

A	constant in Eq. (27)
A_d	outside surface area of a smooth tube with outside diameter, d
A_f	surface area of fin flank
A_r	surface area of inter-fin spacing

B	constant in Eq. (27)
B_f	empirical constant in Eq. (33)
B_{flank1}	empirical constant for pin flank 1
B_{flank2}	empirical constant for pin flank 2
B_l	empirical lead constant in Eq. (37)
B_{root1}	empirical constant for root 1
B_{root2}	empirical constant for root 2
B_s	empirical constant in Eq. (34)
B_t	empirical constant in Eq. (32)
B_{tip}	empirical constant for pin tip
b	fin spacing at fin tip or longitudinal pin spacing at pin tip
c	constant in Eq. (17)
d	outside diameter of plain tube or fin or pin-root diameter of finned or pinned tube
d_o	fin or pin-tip diameter of fin or pin tube
e	fin height of convex profile
F_σ	surface-tension force of condensate
f_f	blanked proportion of the fin flank for unflooded part of the fin tube
f_s	blanked proportion of the inter-fin space for unflooded part of the fin tube
f_x	normalized component of gravity (i.e. in horizontal plane)
g	specific force of gravity
j	number of pins in unflooded region
h	fin or pin height
h_{fg}	specific enthalpy of vapourization
h_v	mean vertical fin or pin height
k	thermal conductivity of condensate
k_w	thermal conductivity of tube wall
L	length of flat plate
L_f	mean vertical fin height over diameter d_o defined by Beatty and Katz [49]
m	$\sqrt{(2\alpha_{\text{flank}}/k_w t)}$

$\dot{m}$	mass flow rate of condensate
Nu_d	average Nusselt number by Honda and Nozu [55] model, defined by Eq. (26)
Nu_{df}	Nusselt number for flooded region, Honda and Nozu [55]
Nu_{du}	Nusselt number for unflooded region, Honda and Nozu [55]
Nu_P	vapour-side Nusselt number for a vertical plate
Nu_T	vapour-side Nusselt number for a horizontal tube
n	total number of pins per circumference
p	fin pitch
Q	total heat-transfer rate through the test tube
q	heat flux on outside of the test tube
q_{flank}	heat flux to fin flank in unflooded part of the tube
q_{int}	heat flux to inter-fin spacing in unflooded part of the tube
q_{plain}	heat flux through plain or smooth tube
q_{tip}	heat flux to fin tip
$q_{\text{tip, flood}}$	heat flux to fin tip in flooded part of the tube
Re	condensate Reynolds number
R_o	fin or pin-tip radius
R_r	fin or pin-root radius
r	radius of curvature of the vapour-liquid interface
r_b	radius of curvature of the vapour-liquid interface at fin bottom
r_t	radius of curvature of the vapour-liquid interface at fin tip
S	distance along the vapour-liquid interface measured from the fin tip
S_m	total fin arc length
s	fin spacing at fin root or longitudinal pin spacing at pin root
s_c	circumferential pin spacing
$\tilde{T}_{wf}$	dimensionless average wall temperatures at fin root in flooded region
$\tilde{T}_{wu}$	dimensionless average wall temperatures at fin root in unflooded region
t	fin-tip thickness or longitudinal pin-tip thickness

t_b	fin-base thickness or longitudinal pin-base thickness
t_c	circumferential pin thickness
V	volume of condensate
We	Weber number
We_c	Weber number for circumferential face of the pin defined by Eq. (44b)
We_l	Weber number for longitudinal face of the pin defined by Eq. (44a)
x	distance along the vapour-liquid interface measured from the fin tip
x_g	characteristic length for gravity-driven flow in Rose [16] model
x_σ	characteristic length for surface-tension-driven flow in Rose [16] model
Y	function of geometric parameters defined by Eq. (45)

Greek Letters

α	mean vapour-side, heat-transfer coefficient
β	angle defined by Eq. (49)
β_m	maximum arc angle
ΔT	temperature difference across the condensate film
δ	condensate film thickness
$\varepsilon_{\Delta T}$	vapour-side, heat-transfer enhancement ratio, heat flux for finned or pinned tube based on fin or pin-root diameter divided by heat flux for smooth tube with the same fin/pin-root diameter, at the same vapour-side, temperature difference
ζ	fin or flute-shape parameter used in Adamek [52] expression
η_f	fin efficiency for flooded region
η_u	fin efficiency for unflooded region
μ	dynamic viscosity of condensate
ξ	active-area enhancement ratio for fin or pin tube
$\xi(\varnothing)$	function given by Eq. (31)
ρ	density of condensate
ρ_v	density of vapour
$\tilde{\rho}$	$\rho-\rho_v$
σ	surface tension

θ fin or pin-tip half angle

$\varnothing$ angle measured from the top of a fin or a pin tube

$\varnothing_f$ condensate flooding or retention angle measured from the top of a fin or a pin tube

Author details

Hafiz Muhammad Ali

Address all correspondence to: hafizmali@kfupm.edu.sa; h.m.ali@uettaxila.edu.pk

1 Center of Research Excellence in Renewable Energy (CoRE-RE), King Fahd University of Petroleum & Minerals (KFUPM), Dhahran, Saudi Arabia

2 Department of Mechanical Engineering, University of Engineering and Technology, Taxila, Pakistan

References

[1] Briggs, A. and Rose, J. W., (1994), *Effect of Fin Efficiency on a Model for Condensation Heat Transfer on a Horizontal Integral Fin Tube,* Int. J. Heat Mass Transfer, Vol. 37, 457–463.

[2] Ali, H. M. and Briggs, A. (2015), *A Semi-Empirical Model for Free-Convection Condensation on Horizontal Pin-Fin Tubes,* Int. J. Heat Mass Transfer, Vol. 81, 157–166.

[3] Nusselt, W., (1916), The Surface Condensation of the Water Vapor, Journal of the Association of German Engineers, Vol. 60, 541–546, 569–575.

[4] Sparrow, E. M. and Gregg, J. L., (1959), *A Boundary-Layer Treatment of Laminar Film Condensation,* Trans. ASME, Vol. 81, 13–18.

[5] Koh, J. C. Y., Sparrow, E. M. and Harnett, J. P., (1961), *The Two Phase Boundary Layer in Laminar Film Condensation,* Int. J. Heat Mass Transfer, Vol. 2, 1, 69–82.

[6] Chen, M. M., (1961), *An Analytical Study of Laminar Film Condensation: Part 1-Flat Plates,* Trans. ASME, Vol. 83, 48–54.

[7] Chen, M. M., (1961), *An Analytical Study of Laminar Film Condensation: Part 2-Single and Multiple Horizontal Tubes,* Trans. ASME, Vol. 83, 55–60.

[8] Rose, J. W., (1988), *Fundamentals of Condensation Heat Transfer: Laminar Film Condensation,* JSME Int. J., Vol. 31, 357–375.

[9] Katz, D. L., Hope, R. E., and Datsko, S. C., (1946), *Liquid Retention on Finned Tubes,* Dept. Eng. Res., Univ. of Michigan, project M592.

[10] Rudy, T. M., and Webb, R. L., (1985), *An Analytical Model to Predict Condensate Retention on Horizontal Integral Fin Tubes,* Trans. ASME, Vol. 107, 361–368.

[11] Rudy, T. M., and Webb, R. L., (1981), *Condensate Retention of Horizontal Integral Fin Tubing,* Adv. Enhanced Heat Transfer, ASME HTD, Vol. 18, 35–41.

[12] Honda, H., Nozu S. and Mitsumori, K., (1983), *Augmentation of Condensation on Finned Tubes by Attaching a Porous Drainage Plate,* Proc. ASME-JSME Therm. Eng. Joint Conf., Vol. 3, 289–295.

[13] Owen, R. G., Sardesai, R. G., Smith, R. A. and Lee, W. C., (1983), *Gravity Controlled Condensation on Low Integral Fin Tubes, in Condenser: Theory and Practice,* 1. Chem. E. Sympos. Serial No. 75, 415–428.

[14] Yau, K. K., Cooper, J. R., and Rose, J. W., (1986), *Horizontal Plain and Low-Finned Condenser Tubes- Effect of Fin Spacing and Drainage Strips on Heat Transfer and Condensate Retention,* Tran. ASME, Vol. 108, 946–950.

[15] Masuda, H. and Rose, J.W., (1987), *Static Configuration of Liquid Films on Horizontal Tubes with Low Radial Fins: Implications for Condensation Heat Transfer,* Proc. Roy. Soc., 410, 125–139.

[16] Rose, J. W., (1994), *An Approximate Equation for the Vapour-side Heat Transfer Coefficient for Condensation on Low Finned Tubes,* Int. J. Heat Mass Transfer, Vol. 37, 865–875.

[17] Wen, X. L., Briggs, A. and Rose, J. W., (1994), *Enhancement of Condensation Heat Transfer on Integral Fin Tubes Using Radius Fin-Root Fillets,* J. Enhanc. Heat Transfer, Vol. 1, 211–217.

[18] Briggs, A., (2005), *Liquid Retention on Three-Dimensional Pin-Fin Tubes,* 2nd Int. Exergy, Energy and Environment Symposium, Kos, Paper No. IEEES2-171.

[19] Ali, H. M. and Briggs, A. (2014), *An investigation of condensate retention on pin-fin tubes,* J. Appl. Therm. Eng, Vol. 63, 503–510.

[20] Ali, H.M and Abubaker, M. (2014), *Effect of Vapour Velocity on Condensate Retention on Horizontal Pin-Fin Tubes,* Energy Convers. Manage. Vol. 86, 1001–1009.

[21] Ali, H.M and Ali, A. (2014), *Measurements and Semi-Empirical Correlation for Condensate Retention on Horizontal Integral-Fin Tubes: Effect of Vapour Velocity,* Appl. Therm. Eng., Vol. 71, Issue 1, 24–33.

[22] Ali, H.M and Abubaker, M. (2015), *Effect of Circumferential Pin Thickness on Condensate Retention as a Function of Vapor Velocity on Horizontal Pin-Fin Tubes,* Appl. Therm. Eng., Vol. 91, 245–251.

[23] Ali, H. M., Ali, H., Ali, M., Imran, S., Kamran, M. S. and Farukh, F., (2016), *Effect of Condensate Flow Rate on Retention Angle on Horizontal Low-Finned Tubes,* Online, Thermal Science. DOI: 10.2298/TSCI151128211A

[24] Yau, K. K., Cooper, J. R., and Rose, J. W., (1985), *Effect of Fin Spacing on the Performance of Horizontal Integral-Fin Condenser Tubes,* Tran. ASME, Vol. 107, 377–383.

[25] Masuda, H. and Rose, J.W., (1985), *An Experimental Study of Condensation of R-113 on Low Integral-Fin Tubes,* Proc. Int. Symp. Heat Transfer., 2, Paper No. 32.

[26] Masuda, H. and Rose, J.W., (1988), *Condensation of Ethylene Glycol on Horizontal Finned Tubes,* Trans. ASME, Vol. 110, 1019–1022.

[27] Wanniarachchi, A. S., Marto, P. J. and Rose, J. W., (1985), *Film Condensation of Steam on Horizontal Finned Tubes: Effect of Fin Spacing, Thickness and Height,* Multiphase Flow Heat Transfer, ASME HTD, Vol. 47, 93–99.

[28] Wanniarachchi, A. S., Marto, P. J. and Rose, J. W., (1986), *Film Condensation of Steam on Horizontal Finned Tubes: Effect of Fin Spacing,* Tran. ASME, Vol. 108, 960–966.

[29] Marto, P. J., Mitrou, E., Wanniarachchi, A. S., and Rose, J. W., (1986), *Film Condensation of Steam on Horizontal Finned Tubes: Effect of Fin Shape,* Proc. 8th Int. Heat Transfer Conf., Vol. 4, 1695–1700.

[30] Marto, P. J., Zebrowski, D., Wanniarachchi, A. S. and Rose, J. W., (1990), *An Experimental Study of R-113 Film Condensation on Horizontal Integral Fin Tubes,* Trans. ASME, Vol. 112, 759–767.

[31] Briggs, A. Wen, X. L. and Rose, J. W., (1992), *Accurate Heat Transfer Measurements for Condensation on Horizontal, Integral-Fin Tubes,* Trans. ASME, Vol. 114, 719–726.

[32] Briggs, A., Song Huang, X. and Rose, J. W., (1995), *An Experimental Investigation of Condensation on Integral Fin Tubes: Effect of Fin Thickness, Height and Thermal Conductivity,* Proc. ASME Nat. Heat Transfer Conf., HTD, Vol. 308, 21–29.

[33] Park, K. J. and Jung, D., (2008), *Optimum Fin Density of Low Fin Tubes for the Condensers of Building Chillers with HCFC123,* J. Energy Conserv. Manage., Vol. 49, 2090–2094.

[34] Sukhatme, S. P., Jagadish, B. S. and Prabhakaran, P., (1990), *Film Condensation of R-11 Vapor on Single Horizontal Enhanced Condenser Tubes,* Trans. ASME, Vol. 112, 229–234.

[35] Briggs, A., Yang, X. X. and Rose, J. W., (1995), *An Evaluation of Various Enhanced Tubes for Shell-Side Condensation of Refrigerant, Heat Transfer in Condensation,* Proc. Eurotherm. Semin. No. 47, Paris, Elsevier Pub. Co., 62–70.

[36] Cheng, W. Y., Wang, C. C., Robert Hu, Y. Z. and Huang, L. W., (1996), *Film Condensation of HCFC-22 on Horizontal Enhanced Tubes,* Int. Comm. Heat Mass Transfer, Vol. 23, 79–90.

[37] Kumar, R., Varma, H. K., Mohanty, B. and Agrawal, K. N., (1998), *Augmentation of Outside Tube Heat Transfer Coefficient during Condensation of Steam over Horizontal Copper Tubes,* Int. Comm. Heat Mass Transfer, Vol. 25, 81–91.

[38] Jung, D. S., Kim, C. B., Cho, S. and Song, K., (1999), *Condensation Heat Transfer Coefficients of Enhanced Tubes with Alternative Refrigerants for CFC11 and CFC12,* Int. J. Refriger., Vol. 22, 548–557.

[39] Kumar, R., Varma, H. K., Mohanty, B. and Agrawal, K. N., (2002), *Augmentation of Heat Transfer During Filmwise Condensation of Steam and R-134a over Single Horizontal Finned Tubes,* Int. J. Heat Mass Transfer, Vol. 45, 201–211.

[40] Kumar, R., Varma, H. K., Mohanty, B. and Agrawal, K. N., (2002), *Prediction of Heat Transfer Coefficient During Condensation of Water and R-134a on Single Horizontal Integral-Fin Tubes*, Int. J. Refriger., Vol. 25, 111–126.

[41] Briggs, A., (2003) *Enhanced Condensation of R-113 and Steam on Three-Dimensional Pin-Fin Tubes*, J. Exp. Heat Transfer, Vol. 16, 61–79.

[42] Baiser, M., and Briggs, A., (2009), *Condensation of Steam on Pin-Fin Tubes: Effect of Circumferential Pin Thickness and Spacing*, J. Heat Transfer Eng., Vol. 30, 1017–1023.

[43] Ali, H. M. and Briggs, A. (2012), *Condensation of R-113 on Pin-Fin Tubes: Effect of Circumferential Pin Thickness and Spacing*, Heat Transfer Eng., Vol. 33, Issue 3, 205–212.

[44] Ali, H. M. and Briggs, A. (2012), *Condensation of Ethylene Glycol on Pin-Fin Tubes: Effect of Circumferential Pin Thickness and Spacing*, Appl. Therm. Eng., Vol. 49, 9–13.

[45] Ali, H. M. and Briggs, A. (2012), *Enhanced Condensation of Ethylene Glycol on Pin-Fin Tubes: Effect of Pin Geometry*, ASME J. Heat Transfer, Vol. 134, 011503.

[46] Ali, H. M. and Briggs, A. (2013), *Condensation Heat Transfer on Pin-Fin Tubes: Effect of Thermal Conductivity and Pin Height*, Appl. Therm. Eng., Vol. 60, 465–471.

[47] Ali, H. M. and Qasim, M. Z. (2015), *Free Convection Condensation of Steam on Horizontal Wire Wrapped Tubes: Effect of Wire Thermal Conductivity, Pitch and Diameter*, Appl. Therm. Eng., Vol. 90, 207–214.

[48] Ali, H. M., Qasim, M. Z. and Ali, M. (2016), *Free Convection Condensation Heat Transfer of Steam on Horizontal Square Wire Wrapped Tubes*, Int. J. Heat Mass Transfer, Vol. 98, 350–358.

[49] Beatty, K. O., and Katz, D. L., (1948), *Condensation of Vapors on Outside of Finned Tubes*, Chem. Eng. Prog., Vol. 44, 55–70.

[50] Briggs, A. and Rose, J. W., (1999), *An Evaluation of Models for Condensation Heat Transfer on Low-Finned Tubes*, J. Enhanc. Heat Transfer, Vol. 6, 51–60.

[51] Gregorig, R., (1954), Film Condensation on Fine-Walled Surfaces, taking into account the Surface Tension, Journal of Applied Mathematics and Physics, Vol. 5, 36–49.

[52] Adamek, T., (1981), Determination of Condensation on Finely Corrugated Surfaces Design of Optimal Wall Profiles, Waerme- and Stoffue-bertragung, Vol. 15, 255–270.

[53] Kedzierski, M.A. and Webb, R.L., (1987), *Experimental Measurements of Condensation on Vertical Plates with Enhanced Fins*, ASME. Boil. Condens. Heat Transfer Equip., HTD, Vol. 85, 87–95.

[54] Rudy, T. M., and Webb, R. L., (1983), *Theoretical Model For Condensation on Horizontal Integral-Fin Tubes*, Am. Inst. Chem. Eng. Symp. Ser., Vol. 79, 11–18.

[55] Honda, H. and Nozu S., (1987), *A Prediction Method for Heat Transfer During Film Condensation on Horizontal Low Integral Fin Tubes*, Trans. ASME, Vol. 109, 218–225.

[56] Cavallini, A., Cesni, G., Del Col, D., Doretti, L., Longo, G. A., Rossetto, L. and Zilio, C., (2003), *Condensation Inside and Outside Smooth and Enhanced Tubes: A Review of Recent Research*, Int. J. Refriger., Vol. 26, 373–392.

[57] Namasivayam, S., (2006), *Condensation on Single Horizontal Integral-Fin Tubes: Effect of Vapour Velocity and Fin Geometry,* PhD Thesis, University of London.

[58] Belghazi, M., Bontemps, A. and Marvillet, C., (2002). *Condensation Heat Transfer on Enhanced Surface Tubes: Experimental Results and Predictive Theory,* Trans. ASME, Vol. 124, 754–761.

Chapter 6

Heat Transfer of Supercritical Fluid Flows and Compressible Flows

Yu Ito

Additional information is available at the end of the chapter

http://dx.doi.org/10.5772/65931

Abstract

In this chapter, the heat transfer between supercritical fluid flows and solid walls and that between compressible flows and solid walls is described. First, the physical fundamentals of supercritical fluids and compressible flows are explained. Second, methods for estimating the heat-transfer performance according to the physical fundamentals and conventional experimental results are described. Then, the known correlations for estimating the heat-transfer performance are introduced. Finally, examples of practical heat exchangers using supercritical fluid flows and/or compressible flows are presented.

Keywords: supercritical fluid flow, compressible flow, nusselt number, reynolds number, mach number, pressure coefficient distribution

1. Introduction

The range of use of heat exchangers is being expanded to extensive applications in various fields. In particular, supercritical fluids and high-speed air, that is,, compressible fluids, are suitable as working fluids.

Supercritical fluid is a phase of substances, in addition to the solid, liquid, and gas phases. In particular, in the vicinity of the critical point, many physical properties behave in an unusual way. For example, the density, viscosity, and thermal conductivity drastically change at the critical point, the specific heat and thermal expansion ratios diverge at the critical point, and the sound velocity is zero at the critical point. The physical properties of a supercritical fluid must be evaluated by the appropriate equation of state and equation of the transport properties.

On the other hand, a compressible flow can be assumed as an ideal gas, but additional dynamic energy, that is, the Mach-number effect, must be considered. Therefore, three types

INTECH open science | open minds

© 2017 The Author(s). Licensee InTech. This chapter is distributed under the terms of the Creative Commons Attribution License (http://creativecommons.org/licenses/by/3.0), which permits unrestricted use, distribution, and reproduction in any medium, provided the original work is properly cited.

of pressures (static, total, and dynamic), four types of temperatures (static, total, dynamic, and recovery), the difference between laminar and turbulent boundary layers, etc., should be distinguished and treated.

2. Equations of state and transport properties of supercritical fluid

Figure 1 shows the *P–T* diagram of a pure substance (water in this case), which is also called a phase diagram. The sublimation curve divides the solid and gas phases, the melting curve divides the solid and liquid phases, and the vaporization curve divides the liquid and gas phases. Two phases coexist on these three curves. When the pressure and/or temperature change across these three coexistence curves of solid-gas, solid-liquid, and liquid-gas, the density discontinuously changes. These three coexistence curves meet at the triple point, which is the unique point where solid, liquid, and gas coexist in equilibrium.

The vaporization curve ends at the critical point. On the vaporization curve, liquid is called the saturation liquid, and gas is called the saturation gas (vapor). When approaching the critical point along the vaporization curve, the density of the saturation liquid decreases, and the density of the saturation gas (vapor) increases. Finally, they meet at the critical point. Fluid overtaking the critical point in temperature and pressure is called the "supercritical fluid."

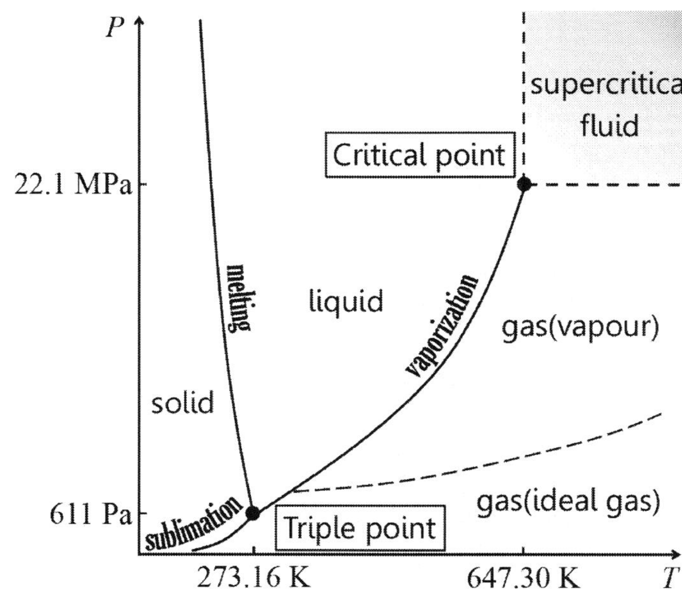

Figure 1. Phase chart on P-T diagram (for water).

Solid	Diffusivity << condensability
Liquid	Diffusivity < condensability
Supercritical fluid	Diffusivity ≈ condensability
Gas (vapor)	Diffusivity > condensability
Gas (ideal gas)	Diffusivity >> condensability

The phase is thermodynamically determined by the Gibbs free energy G:

$$G = H - TS = U - TS + PV \tag{1}$$

Where H is the enthalpy, S is the entropy, U is the internal energy, and V is the specific volume [1].

That is, the phase is determined by the balance between the diffusivity caused by the thermal mobility of the molecules and the condensability by intermolecular forces. The diffusivity caused by thermal mobility increases with the temperature. The condensability by intermolecular forces increases with the density. In general, the following relationships hold:

In **Figure 1**, the first-order differentials of the Gibbs free energy

$$dG = -SdT + VdP \tag{2}$$

$$\left(\frac{dG}{dT}\right)_{\mathrm{P}} = -S \tag{3}$$

$$\left(\frac{dG}{dP}\right)_{\mathrm{T}} = V = \frac{1}{\rho} \tag{4}$$

are discontinuous across the three coexistence curves, but the first-order differentials of the Gibbs free energy are continuous at the critical point. In addition, the second-order differentials of the Gibbs free energy are discontinuous at the critical point.

$$\left(\frac{d^2G}{dT^2}\right)_{\mathrm{P}} = -\frac{1}{T}\left(\frac{dH}{dT}\right)_{\mathrm{P}} = -\frac{1}{T}C_{\mathrm{P}} \tag{5}$$

$$\left(\frac{d^2G}{dP^2}\right)_{\mathrm{T}} = \left(\frac{dV}{dP}\right)_{\mathrm{T}} = -VK_{\mathrm{T}} \tag{6}$$

Here, ρ is the density, C_{P} is the isobaric specific heat, and K_{T} is the isothermal compressibility. At the critical point, the density drastically changes, the specific heat and thermal expansion ratio diverge, and the sound velocity is zero.

Figures 2 and **3** show the isobaric and isothermal changes of the density, viscosity, and kinematic viscosity by using the data from references [2, 3]. Both the density (derived from

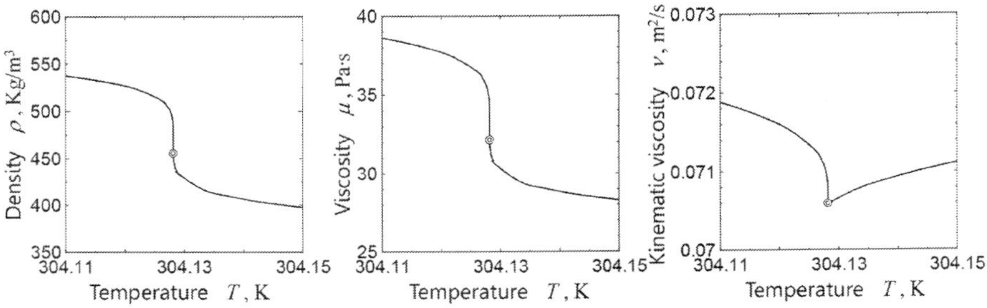

Figure 2. Isobaric changes of the density, viscosity, and kinematic viscosity near the critical point where $T_{critical}$ = 304.1282 K and $P_{critical}$ = 7.3773 MPa (for carbon dioxide).

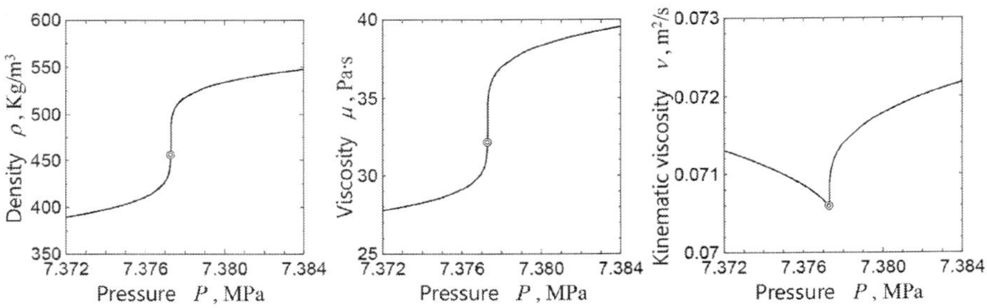

Figure 3. Isothermal changes of the density, viscosity, and kinematic viscosity near the critical point where $T_{critical}$ = 304.1282 K and $P_{critical}$ = 7.3773 MPa (for carbon dioxide).

the equation of state) and the viscosity (derived from the equation of the transport properties) drastically change at the critical point, and the derivatives with respect to temperature and pressure diverge at the critical point. The kinematic viscosity (combined with the density and viscosity) has an extremum value at the critical point. The equations of state and the transport properties should consider these types of tricky features in the vicinity of the critical point for transcritical- and supercritical-fluid flows.

The most important substances in practical applications are carbon dioxide and water, although all substances have a supercritical-fluid phase. Recently, accurate correlations for the equations of state and the transport properties containing the critical point have been proposed.

For carbon dioxide, Span and Wagner proposed the equation of state from the triple point to 1100 K at pressures up to 800 MPa [2]. Their equation of state is briefly introduced here. They expressed the fundamental equation in the form of the Helmholtz energy A:

$$A = U - TS = H - RT - TS \tag{7}$$

with two independent variables—the density ρ and temperature T. The dimensionless Helmholtz energy $\phi = A/RT$ is divided into a part obeying the ideal gas behavior ϕ° and a part that deviates from the ideal gas behavior ϕ^{r} [2]:

$$\phi(\delta,\tau) = \phi^{\circ}(\delta,\tau) + \phi^{r}(\delta,\tau), \tag{8}$$

Where $\delta = \rho/\rho_c$ is the reduced density; $\tau = T_c/T$ is the inverse reduced temperature; and ρ_c and T_c are the density and temperature, respectively, at the critical point. Then, all of the other thermodynamic properties can be obtained by the combined derivatives of Eq. (7) using the Maxwell relations [1].

Pressure

$$P(T,\rho) = -\left(\frac{\partial A}{\partial V}\right)_T \text{then} \frac{P(\delta,\tau)}{\rho RT} = 1 + \delta\phi_\delta^r \tag{9}$$

Entropy

$$S(T,\rho) = -\left(\frac{\partial A}{\partial T}\right)_V \text{then} \frac{S(\delta,\tau)}{R} = \tau[\phi_\tau^{\circ} + \phi_\tau^r] - \phi^{\circ} - \phi^r \tag{10}$$

Internal energy

$$U(T,\rho) = A - T\left(\frac{\partial A}{\partial T}\right)_V \text{then} \frac{U(\delta,\tau)}{RT} = \tau[\phi_\tau^{\circ} + \phi_\tau^r] \tag{11}$$

Isochoric specific heat

$$C_V(T,\rho) = \left(\frac{\partial U}{\partial T}\right)_V \text{then} \frac{C_V(\delta,\tau)}{R} = -\tau^2[\phi_{\tau\tau}^{\circ} + \phi_{\tau\tau}^r] \tag{12}$$

Enthalpy

$$H(T,\rho) = A - T\left(\frac{\partial A}{\partial T}\right)_V - V\left(\frac{\partial A}{\partial V}\right)_T \text{then} \frac{H(\delta,\tau)}{RT} = 1 + \tau[\phi_\tau^{\circ} + \phi_\tau^r] + \delta\phi_\delta^r \tag{13}$$

Isobaric specific heat

$$C_P(T,\rho) = \left(\frac{\partial H}{\partial T}\right)_P \text{then} \frac{C_P(\delta,\tau)}{R} = -\tau^2[\phi_{\tau\tau}^{\circ} + \phi_{\tau\tau}^r] + \frac{[1 + \delta\phi_\delta^r - \delta\tau\phi_{\delta\tau}^r]^2}{1 + 2\delta\phi_\delta^r + \delta^2\phi_{\delta\delta}^r} \tag{14}$$

Saturated specific heat

$$C_\sigma(T) = \left(\frac{\partial H}{\partial T}\right)_P + T\left(\frac{\partial P}{\partial T}\right)_V \left(\frac{\partial P_{sat}}{\partial T}\right) / \left(\frac{\partial P}{\partial V}\right)_{T_{sat}} \tag{15}$$

then,

$$\frac{C_\sigma(\delta,\tau)}{R} = -\tau^2[\phi_{\tau\tau}^{\circ} + \phi_{\tau\tau}^r] + \frac{1 + \delta\phi_\delta^r - \delta\tau\phi_{\delta\tau}^r}{1 + 2\delta\phi_\delta^r + \delta^2\phi_{\delta\delta}^r}\left[\{1 + \delta\phi_\delta^r - \delta\tau\phi_{\delta\tau}^r\} - \frac{\rho_c}{R\delta}\frac{dP_{sat}}{dT}\right] \tag{16}$$

Speed of sound

$$w(T,\rho) = \sqrt{\left(\frac{\partial P}{\partial \rho}\right)_S} \text{ then} \frac{w^2(\delta,\tau)}{RT} = 1 + 2\delta\phi_\delta^r + \delta^2\phi_{\delta\delta}^r - \frac{[1+\delta\phi_\delta^r - \delta\tau\phi_{\delta\tau}^r]^2}{\tau^2[\phi_{\tau\tau}^\circ + \phi_{\tau\tau}^r]} \tag{17}$$

etc.

Here,

$$\phi_\delta = \left(\frac{\partial\phi}{\partial\delta}\right)_\tau, \ \phi_{\delta\delta} = \left(\frac{\partial^2\phi}{\partial\delta^2}\right)_\tau, \ \phi_\tau = \left(\frac{\partial\phi}{\partial\tau}\right)_\delta, \ \phi_{\tau\tau} = \left(\frac{\partial^2\phi}{\partial\tau^2}\right)_\delta \text{ and } \phi_{\delta\tau} = \left(\frac{\partial^2\phi}{\partial\delta\partial\tau}\right).$$

For carbon dioxide, Vesovic et al. proposed transport properties in the temperature range of 200–1500 K for the viscosity μ and in the temperature range of 200–1000 K for the thermal conductivity k [3]. Their equations of the transport properties μ and k are briefly introduced. Their fundamental equation combines three independent parts: a part obeying the ideal gas behavior $\mu^\circ(T)$ and $k^\circ(T)$, a part with excess properties because of the elevated density $\Delta\mu(\rho,T)$ and $\Delta k(\rho,T)$, and a part with an enhancement in the vicinity of the critical point $\Delta_c\mu(T)$ and $\Delta_c k(T)$:

$$\mu(\rho,T) = \mu^\circ(T) + \Delta\mu(\rho,T) + \Delta_c\mu(T) \tag{18}$$

$$k(\rho,T) = k^\circ(T) + \Delta k(\rho,T) + \Delta_c k(T) \tag{19}$$

For water, Wagner and Pruß proposed the equation of state for the temperature range of 251.2–1273 K and pressures up to 1000 MPa [4]. Huber et al. proposed the transport properties from the melting temperature to 1173 K at 1000 MPa [5, 6].

3. Heat transfers between supercritical fluid flow and solid

As mentioned in Section 2, the kinematic viscosity of a supercritical fluid is less than those of a liquid and gas; therefore, the Reynolds number, Re, of a supercritical fluid flow is higher than those of a liquid and gas flow with the same velocity, and a turbulent flow is easily formed. For heat transfer in a turbulent flow, Dittus and Boelter proposed a correlation of the Nusselt number using the Re and Prandtl number, Pr, for a liquid flow in a circular automobile radiator [7] as shown in **Figure 4**.

$$\mathrm{Nu}_{\mathrm{local,turb}} = 0.023\mathrm{Re}_{\mathrm{local}}^{0.8}\mathrm{Pr}_{\mathrm{local}}^{n} \tag{20}$$

$$\mathrm{Re}_{\mathrm{local}} = \frac{u_{\mathrm{local}}D}{\nu_{\mathrm{local}}} = \frac{\rho_{\mathrm{local}}u_{\mathrm{local}}D}{\mu_{\mathrm{local}}} = \frac{4m}{\pi D\mu_{\mathrm{local}}} \tag{21}$$

$$\mathrm{Pr}_{\mathrm{local}} = \frac{\nu_{\mathrm{local}}}{\kappa_{\mathrm{local}}} = \frac{\mu_{\mathrm{local}}C_{P,\mathrm{local}}}{k_{\mathrm{local}}} \tag{22}$$

Here, the superscript $n = 0.3$ for $T_{\mathrm{wall}} < T_{\mathrm{fluid}}$ or $n = 0.4$ for $T_{\mathrm{wall}} > T_{\mathrm{fluid}}$, u_{local} is the average velocity across the cross section, D is the diameter of the tube, μ is the viscosity, m is the mass

http://dx.doi.org/10.5772/65931

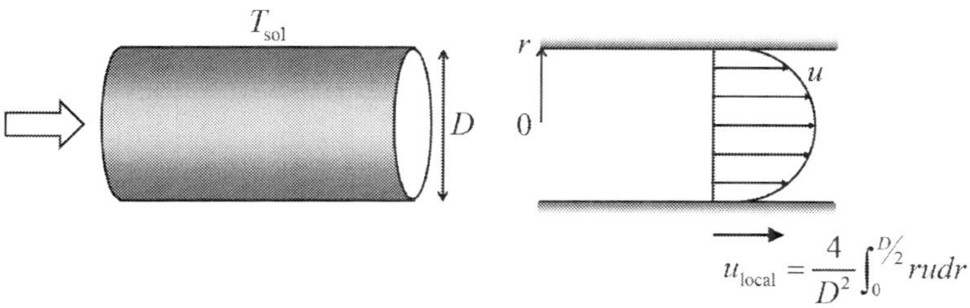

Figure 4. Heat transfer between a supercritical fluid flow and a circular solid tube wall.

flow rate, κ is the thermal diffusivity, and k is the heat conductivity. For liquid and gas flows, the fluid properties and flow conditions can be regarded as constant throughout the entire region in most practical cases because the fluid properties are insensitive to temperature and pressure changes in the tube. Therefore, the inlet values of the physical properties and flow conditions can be used, and Re and Pr can be regarded as constant throughout the entire tube. On the other hand, the fluid properties of a supercritical fluid are very sensitive to temperature and pressure changes in the tube. Thus, in the tube, the density gradually changes because of the heat input and/or pressure loss, the local average velocity changes, and even Re and Pr change. Unfortunately, the Dittus-Boelter correlation with the inlet values of the physical properties and flow conditions cannot be directly used for heat transfer in a supercritical turbulent flow. Liao and Zhao measured the rate of the heat transfer between a supercritical carbon dioxide flow and a circular solid tube wall for $T_{wall} < T_{fluid}$ [8]. Their tube was set in the horizontal direction. They proposed a correlation of area-averaged Nusselt numbers as functions of the Reynolds and Prandtl numbers defined at the temperatures of the mean bulk and the wall.

$$\mathrm{Nu}_{ave,turb} = 0.128\mathrm{Re}_{wall}^{0.8}\mathrm{Pr}_{wall}^{0.3}\left[\frac{\mathrm{Gr}}{\mathrm{Re}_{bulk}^2}\right]^{0.205}\left[\frac{\rho_{bulk}}{\rho_{wall}}\right]^{0.437}\left[\frac{C_{P,bulk}}{C_{P,wall}}\right]^{0.411} \mathrm{for CO_2} \quad (23)$$

$$\mathrm{Gr} = \frac{[\rho_{wall} - \rho_{bulk}]\rho_{bulk} g D^3}{\mu_{bulk}^2} \quad (24)$$

$$\mathrm{Re}_{bulk} = \frac{4m}{\pi D \mu(T_{bulk}, P_{bulk})} \quad (25)$$

$$\rho_{bulk} = \rho(T_{bulk}, P_{bulk}) \quad (26)$$

$$\mu_{bulk} = \mu(T_{bulk}, P_{bulk}) \quad (27)$$

$$\mathrm{Re}_{wall} = \frac{4m}{\pi D \mu(T_{wall}, P_{wall})} \quad (28)$$

$$\mathrm{Pr}_{wall} = \frac{\nu(T_{wall}, P_{wall})}{\kappa(T_{wall}, P_{wall})} = \frac{\mu(T_{wall}, P_{wall}) C_P(T_{wall}, P_{wall})}{k(T_{wall}, P_{wall})} \quad (29)$$

$$\rho_{\text{wall}} = \rho(T_{\text{wall}}, P_{\text{wall}}) \tag{30}$$

$$C_{\text{P,wall}} = C_{\text{P}}(T_{\text{wall}}, P_{\text{wall}}) \tag{31}$$

$$C_{\text{P,bulk}} = C_{\text{P}}(T_{\text{bulk}}, P_{\text{bulk}}) \tag{32}$$

Here, $T_{\text{bulk}} = [T_{\text{in}} + T_{\text{out}}]/2$ and T_{wall} are constant. This correlation is applicable in the range of 7.4 MPa $< P_{\text{bulk}} <$ 12.0 MPa, 20°C $< T_{\text{bulk}} <$ 110°C, 2°C $< T_{\text{bulk}} - T_{\text{wall}} <$ 30°C, 0.02 kg/min $< \dot{m} <$ 0.2 kg/min, 1025 $< \text{Gr}/\text{Re}_{\text{bulk}}^2 <$ 1022 for the horizontal long tubes of 0.50 mm $< d <$ 2.16 mm.$\text{Re}_{\text{wall}}^{0.8}$ and $\text{Pr}_{\text{wall}}^{0.3}$ were originally derived from the Dittus-Boelter correlation. $\text{Gr}/\text{Re}_{\text{bulk}}^2$ is the effect of buoyancy in the radial direction of a horizontal tube. The density of fluid at a temperature and pressure in the vicinity of the critical point is very sensitive to changes in temperature; thus, the effect of the buoyancy derived from the temperature difference between the bulk and wall cannot be ignored. This effect is enhanced as the diameter of the tube increases.

Ito et al. proposed an airfoil heat exchanger, which is applied between a compressible airflow and a liquid or a supercritical fluid flow [9]. It has an outer airfoil shape suitable for high-speed airflow and contains several tubes for a high-pressure liquid or a supercritical fluid flow. The researchers installed a cascade of airfoil heat exchangers into a subsonic wind tunnel at a temperature of T_{air} and measured the heat-transfer coefficient of a liquid or a supercritical fluid flow at a temperature of $T_{\text{scf}} < T_{\text{air}}$ in a vertical tube. They derived correlations for supercritical carbon dioxide and compressed water at a pressure of $P_{\text{scf}} \leq 30$ MPa as follows:

$$\text{Nu}_{\text{ave,turb}} = \begin{cases} 0.0230\text{Re}^{0.808}\text{Pr}^{0.300} & \text{for} H_2O \\ 0.0231\text{Re}^{0.823}\text{Pr}^{0.300} & \text{for} CO_2 \end{cases} \tag{33}$$

These correlations are very simple and similar to the Dittus-Boelter correlation in Eq. (20) but have sufficient accuracy. Ito et al. used accurate equations of state and the transport properties, as mentioned in Section 2. They said in reference [9] that ordinary correlations (of course, containing the Dittus-Boelter correlation) for liquid and gas can be used when sufficiently accurate equations of state and the transport properties are used. However, the physical properties at a temperature and pressure in the vicinity of the critical point continuously change throughout the tube because of the heat input and/or pressure loss; therefore, changes in these physical properties throughout the tube should be sufficiently considered. For example, the present author recommends the numerical integration of local heat transfer correlations using local accurate physical properties for the entire tube.

4. Thermofluid dynamics of compressible flow on solid wall

4.1. Meanings of temperature and pressure of compressible flow

A stationary fluid pressure of P [Pa], specific volume of V [m^3/kg], and constant temperature T stores a mechanical energy of e_{pre} [J/kg]. Here,

http://dx.doi.org/10.5772/65931

$$e_{\mathrm{pre}} = PV. \tag{34}$$

The "pressure" (often called "static pressure") P is the potential of the mechanical energy level contained in a stationary fluid. A motional fluid has an additional dynamic energy e_{dyn} [J/kg]:

$$e_{\mathrm{dyn}} = \frac{1}{2}u^2. \tag{35}$$

in addition to e_{pre}; therefore,

$$e_{\mathrm{pre}} + e_{\mathrm{dyn}} = PV + \frac{1}{2}u^2 = V[P + P_{\mathrm{dyn}}] = VP_{\mathrm{tot}} = e_{\mathrm{mech}}. \tag{36}$$

$$P_{\mathrm{dyn}} = \frac{1}{2V}u^2 = \frac{1}{2}\rho u^2. \tag{37}$$

$$P_{\mathrm{tot}} = P + P_{\mathrm{dyn}}. \tag{38}$$

Here, P_{dyn} [Pa] is called "dynamic pressure" and is an index of the dynamic mechanical energy level contained in a motional fluid. Further, e_{mech} [J/kg] is called "total mechanical energy." Moreover, P_{tot} [Pa] is called the "total pressure" and is an index of the total mechanical energy level contained in a motional fluid. Some processes are reversible between mechanical energies of e_{pre} and e_{dyn} in cases where e_{pre} and e_{dyn} transform in the equilibrium processes. For example, using a nozzle, P_{dyn} increases, P decreases, and P_{tot} is constant in an acceleration section, and P_{dyn} decreases, P increases, and P_{tot} is constant in a deceleration section. Some process are irreversible between e_{pre} and e_{dyn} in cases where e_{pre} and e_{dyn} transform in nonequilibrium processes. For example, because of friction, P remains constant, P_{dyn} decreases, and P_{tot} also decreases.

Next, we consider thermal energy. A stationary fluid at an isochoric specific heat of C_V [J/(kg K)] stores a relative internal energy of e [J/kg] from e_0 at the standard temperature T_0:

$$e - e_0 = \int_{T_0}^{T} C_V dT. \tag{39}$$

Here, the internal energy is an index of the thermal energy level contained in a stationary fluid. In the case of a constant C_V,

$$e - e_0 = C_V[T - T_0]. \tag{40}$$

The "temperature" (often called "static temperature") T is an index of the energy level contained in a stationary fluid.

In cases where a fluid is assumed as an ideal gas,

$$PV = RT \Leftrightarrow P = \rho RT, \tag{41}$$

where R is the gas constant, and ρ is the density, which is equal to $1/V$. Then,

$$R = \frac{Pv}{T} = \frac{P_0 v_0}{T_0} \tag{42}$$

A stationary fluid at an isochoric specific heat of C_V [J/(kg K)] stores a relative enthalpy of h [J/kg] from h_0 at the standard temperature of T_0, a pressure of P_0, and a specific volume of V_0. Videlicet, enthalpy is a combination of internal energy and mechanical energy. Here,

$$h\text{–}h_0 = [e + PV]\text{–}[e_0 + P_0 V_0] = \int_{T_0}^{T} C_V dT + \int_{T_0}^{T} R dT. \tag{43}$$

In the case of a constant C_V,

$$h\text{–}h_0 = C_V[T\text{–}T_0] + R[T\text{–}T_0] = C_P[T\text{–}T_0], \tag{44}$$

$$C_P = C_V + R, \tag{45}$$

where C_P is the isobaric specific heat. A motional fluid has an additional dynamic energy e_{dyn} [J/kg], as shown in Eq. (35). If a motional fluid suddenly stops, dynamic energy can be converted into enthalpy. Then, the following equation applies:

$$h\text{–}h_0 + e_{\text{dyn}} = C_P[T\text{–}T_0] + \frac{1}{2}u^2 = C_P[T + T_{\text{dyn}}\text{–}T_0] = C_P[T_{\text{tot}}\text{–}T_0]. \tag{46}$$

$$T_{\text{dyn}} = \frac{1}{2C_P}u^2. \tag{47}$$

$$T_{\text{tot}} = T + T_{\text{dyn}}. \tag{48}$$

Here, T_{dyn} [K] is called the "dynamic temperature" and is an index of the dynamic energy level contained in a motional fluid. Moreover, T_{tot} [K] is called the "total temperature" and is an index of the total energy level contained in a motional fluid. Some processes are reversible between mechanical energies of h and e_{dyn} in cases where e_{dyn} transforms into PV in equilibrium processes. For example, using a nozzle, in an acceleration section, T_{dyn} increases, h decreases, T_{tot} is constant, and vice versa. Some processes are irreversible in cases where e_{dyn} transforms into $C_V T$ in nonequilibrium processes. For example, because of friction, T_{dyn} decreases, T increases, and T_{tot} remains constant; however, T cannot be converted into T_{dyn} again.

http://dx.doi.org/10.5772/65931

4.2. Isentropic change and sound speed of ideal gas

The specific heat ratio γ is defined as:

$$\gamma = \frac{C_P}{C_V} \tag{49}$$

From Eqs. (45) and (49),

$$C_V = \frac{R}{\gamma - 1}, \; C_P = \frac{\gamma R}{\gamma - 1} \tag{50}$$

This equation is substituted into Eqs. (40) and (44). Then,

$$e - e_0 = \frac{R}{\gamma - 1}[T - T_0], \;\; h - h_0 = \frac{\gamma R}{\gamma - 1}[T - T_0] \tag{51}$$

$$de = \frac{R}{\gamma - 1} dT, dh = \frac{\gamma R}{\gamma - 1} dT \tag{52}$$

The change in the entropy ds is defined as:

$$Tds = de + pdV, Tds = dh - VdP \tag{53}$$

$$ds = \frac{de + pdV}{T}, ds = \frac{dh - VdP}{T} \tag{54}$$

When isentropic change $ds = 0$,

$$0 = ds = C_V \frac{dT}{T} + R\frac{dV}{V}, 0 = ds = C_P \frac{dT}{T} - R\frac{dP}{P} \tag{55}$$

$$0 = \frac{R}{\gamma - 1}\frac{dT}{T} - R\frac{d\rho}{\rho}, 0 = \frac{\gamma R}{\gamma - 1}\frac{dT}{T} - R\frac{dP}{P} \tag{56}$$

$$\frac{1}{\gamma - 1}\frac{dT}{T} = \frac{d\rho}{\rho}, \; \frac{\gamma}{\gamma - 1}\frac{dT}{T} = \frac{dP}{P} \tag{57}$$

We totally differentiate Eq. (41), obtaining the following:

$$dP = RTd\rho + \rho TdR + \rho RdT \tag{58}$$

$$\frac{dP}{P} = \frac{d\rho}{\rho} + \frac{dR}{R} + \frac{dT}{T} = \frac{d\rho}{\rho} + \frac{dT}{T} \Leftrightarrow \frac{dT}{T} = \frac{dP}{P} - \frac{d\rho}{\rho} \tag{59}$$

We substitute the final equation of Eq. (59) and Eq. (45) into the rightmost part of Eq. (55):

$$0 = ds = C_P\frac{dT}{T} - R\frac{dP}{P} = C_P\frac{dP}{P} - C_P\frac{d\rho}{\rho} - R\frac{dP}{P} = C_V\frac{dP}{P} - C_P\frac{d\rho}{\rho} \tag{60}$$

$$0 = \frac{R}{\gamma - 1}\frac{dP}{P} - \frac{\gamma R}{\gamma - 1}\frac{d\rho}{\rho} \tag{61}$$

$$\gamma\frac{d\rho}{\rho} = \frac{dP}{P} \tag{62}$$

$$\frac{dP}{d\rho} = \gamma\frac{P}{\rho} \tag{63}$$

We integrate Eqs. (57) and (62):

$$\frac{T}{\rho^{\gamma-1}} = \text{const}, \quad \frac{T}{P^{\frac{\gamma-1}{\gamma}}} = \text{const}, \quad \frac{P}{\rho^{\gamma}} = \text{const} \tag{64}$$

The sound speed a is defined as:

$$a^2 = \left[\frac{dP}{d\rho}\right]_S \tag{65}$$

Eqs. (63) and (41) are substituted into Eq. (65), yielding the following:

$$a^2 = \gamma\frac{P}{\rho} = \gamma RT \tag{66}$$

4.3. Relationships of static and total values in isentropic compressible flow

The one-dimensional energy equation of an isentropic flow at an arbitrary cross section is derived by using Eq. (46) as:

$$h + \frac{1}{2}u^2 = \text{const} \tag{67}$$

When the enthalpy and velocity are h_1 and u_1 at an arbitrary cross section 1,

$$h_1 + \frac{1}{2}u_1^2 = h + \frac{1}{2}u^2 \tag{68}$$

This relationship is true even if cross section 1 corresponds to the stagnant cross section 0 (h_0 and $u_0 = 0$); therefore,

$$h_0 = h + \frac{1}{2}u^2 \tag{69}$$

Eqs. (44) and (50) are substituted into Eq. (69), yielding the following:

$$\frac{\gamma RT_0}{\gamma-1} = \frac{\gamma RT}{\gamma-1} + \frac{1}{2}u^2 \tag{70}$$

Eq. (66) is substituted into Eq. (70):

$$\frac{a_0^2}{\gamma-1} = \frac{a^2}{\gamma-1} + \frac{1}{2}u^2 \tag{71}$$

We multiply by $\frac{\gamma-1}{a^2}$ and substitute Eq. (66), obtaining the following:

$$\frac{a_0^2}{a^2} = \frac{RT_0}{RT} = \frac{T_0}{T} = 1 + \frac{\gamma-1}{2}\frac{u^2}{a^2} \tag{72}$$

At the stagnant cross section 0, the static temperature T_0 is equal to the total temperature T_{tot}; therefore,

$$\frac{T_{tot}}{T} = 1 + \frac{\gamma-1}{2}\mathrm{M}^2, \; T = \frac{T_{tot}}{1+\frac{\gamma-1}{2}\mathrm{M}^2}, \tag{73}$$

where M is the local Mach number. From Eqs. (64) and (73),

$$\frac{P_{tot}}{P} = \frac{T_{tot}^{\frac{\gamma}{\gamma-1}}}{T^{\frac{\gamma}{\gamma-1}}} = \left[\frac{T_{tot}}{T}\right]^{\frac{\gamma}{\gamma-1}} = \left[1+\frac{\gamma-1}{2}\mathrm{M}^2\right]^{\frac{\gamma}{\gamma-1}}, P = \frac{P_{tot}}{\left[1+\frac{\gamma-1}{2}\mathrm{M}^2\right]^{\frac{\gamma}{\gamma-1}}} \tag{74}$$

4.4. Relationships of local Mach number, pressure and temperature of flows on adiabatic walls

Figure 5 shows the pressure distribution on a plane and an airfoil. On both the plane and the airfoil, boundary layers are formed. The pressure $P_{local,bound}$ in a boundary layer is almost equal to the pressure $P_{local,main}$ in a main flow outside of the boundary layer; therefore, the pressure in a boundary layer can be expressed by using relationships of the isentropic main flow. That is, $P_{local,bound} = P_{local,main}$. Afterwards, both the pressures are expressed as P_{local}.

The pressure distribution on a solid wall is usually expressed by pressure coefficient S_{local}, which is defined as:

$$S_{local} = \frac{P_{tot,in}-P_{local}}{\frac{1}{2}\rho u_{in}^2}, P_{local} = P_{tot,in}-\frac{1}{2}\rho u_{in}^2 S_{local} \tag{75}$$

but is sometimes expressed by another pressure coefficient η_{local}, which is defined as:

$$\eta_{local} = \frac{P_{in}-P_{local}}{\frac{1}{2}\rho u_{in}^2}, P_{local} = P_{in}-\frac{1}{2}\rho u_{in}^2 \eta_{local}. \tag{76}$$

The two expressions are related as follows:

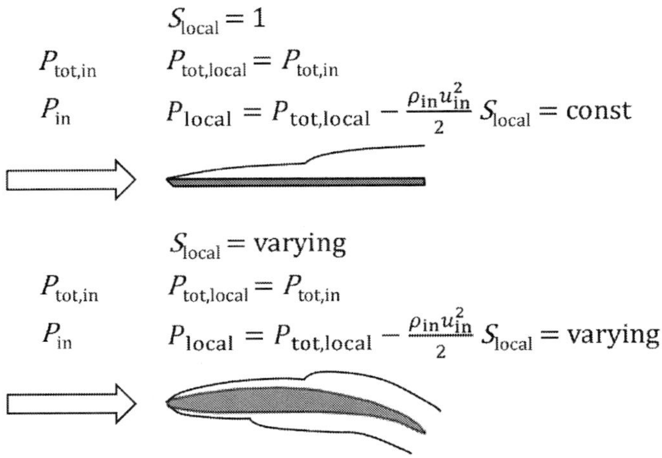

Figure 5. Pressure distributions of flows on a plane and an airfoil.

$$S_{local} = \frac{P_{tot,in} - P_{local}}{\frac{1}{2}\rho u_{in}^2} = \frac{P_{tot,in} - P_{in} + P_{in} - P_{local}}{\frac{1}{2}\rho u_{in}^2} = \frac{\frac{1}{2}\rho u_{in}^2 + P_{in} - P_{local}}{\frac{1}{2}\rho u_{in}^2} = 1 + \eta_{local}. \tag{77}$$

On a plane, S_{local} is unity everywhere; thus, P_{local} is constant everywhere. On the other hand, on an airfoil, S_{local} varies with the location; thus, P_{local} varies.

Figure 6 shows the temperature distribution on an adiabatic plane and an airfoil. In flows on an adiabatic wall, the total temperature $T_{tot,local}$ remains constant at the inlet total temperature $T_{tot,in}$. For incompressible flows, that is, with the Mach number $M_{local} = 0$, the static temperature T_{local} is always the same as $T_{tot,local}$. Then, T_{local} remains constant everywhere on both an adiabatic plane and an airfoil. On the other hand, for compressible flows, the Mach number M_{local} varies according to the following equation, which is derived from Eq. (74):

$$M_{local} = \sqrt{\frac{2}{\gamma - 1}\left\{\left[\frac{P_{tot,local}}{P_{local}}\right]^{\frac{\gamma-1}{\gamma}} - 1\right\}} \tag{78}$$

Here, the static temperature T_{local} varies with respect to M_{local} for compressible flows.

$$T_{local} = \frac{T_{tot,local}}{1 + \frac{\gamma - 1}{2} M_{local}^2} \tag{79}$$

On an adiabatic plane, M_{local} is constant. Thus, T_{local} remains constant anywhere on an adiabatic plane, even in cases of compressible flows. On the other hand, on an adiabatic airfoil, M_{local} varies with the location; therefore, T_{local} varies in cases of compressible flows.

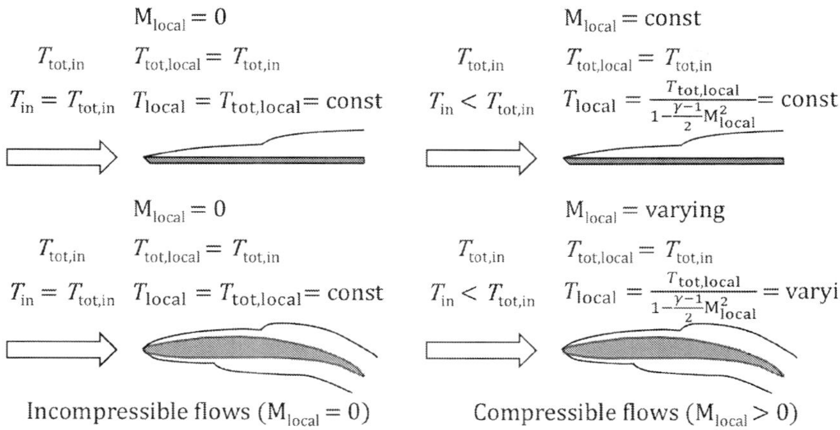

Figure 6. Temperature distributions of flows on an adiabatic plane and an airfoil.

4.5. Recovery temperature definition in boundary layer in compressible flow on adiabatic, heating and cooling walls

Eckert surveyed and organized the heat transfer in a boundary layer in a compressible flow on a wall [10]. In a boundary layer on an adiabatic plane, the adiabatic-wall temperature reaches T_r. This is called the "recovery temperature." **Figure 7** shows a schematic of the total temperature T_{tot} and static temperature T profiles, as well as the recovery temperature T_r in the vicinity of an adiabatic solid surface with a boundary layer in a compressible flow. As described in Section 4.1, a compressible flow has a measurable dynamic energy; then, the static temperature T in a boundary layer increases because of the braking effect, which converts a dynamic energy to a thermal energy. At the same time, heat generated by the braking effect conducts to the outside of the boundary layer. Therefore, the static temperature T and total temperature T_{tot} in the boundary layer approach the recovery temperature T_r on the wall, as shown in the middle in **Figure 7**.

In cases where a thermal boundary layer is completely inside a momentum boundary layer, that is, $\mathrm{Pr} \geq 1$ the heat generated by the braking effect uses the rise of the static temperature T. The recovery temperature T_r on an adiabatic wall is equal to the total temperature $T_{tot,main}$ of the main flow outside of the boundary layer. On the other hand, in cases where a thermal boundary layer protrudes from the edge of a momentum boundary layer, that is, $\mathrm{Pr} < 1$, only part of the heat generated by the braking effect uses the rise of the static temperature T; then, the recovery temperature T_r on an adiabatic wall has an intermediate value between the total temperature $T_{tot,main}$ and the static temperature T_{main} of the main flow outside of the boundary layer. Eckert proposed an equation for the local recovery temperature $T_{r,local}$ on an adiabatic wall.

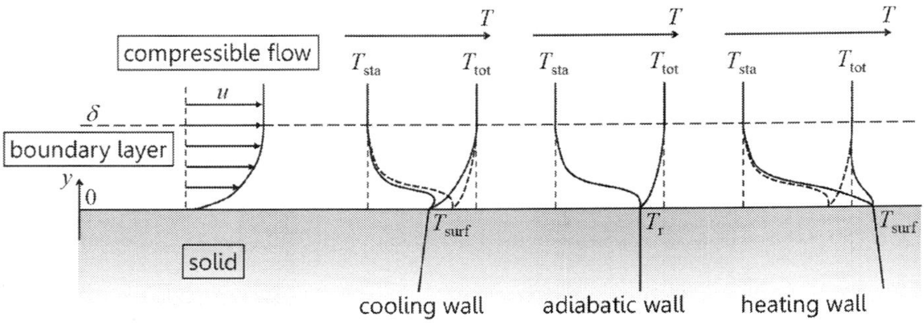

Figure 7. Total-, static-, and recovery-temperature profiles in the vicinity of cooling, adiabatic, and heating solid surfaces with a boundary layer in a compressible flow.

$$T_{\mathrm{r,local}} = T_{\mathrm{main}} + [T_{\mathrm{tot,main}} - T_{\mathrm{main}}]r_{\mathrm{local}} = T_{\mathrm{main}} + T_{\mathrm{dyn,main}} r_{\mathrm{local}} \tag{80}$$

$$r_{\mathrm{local}} = \begin{cases} \min(1, \mathrm{Pr}^{1/2}) & \text{for a laminar boundary layer} \\ \min(1, \mathrm{Pr}^{1/3}) & \text{for a turbulent boundary layer} \end{cases} \tag{81}$$

Here, r_{local} is the "temperature recovery factor," which is the ratio of the recovery temperature to the dynamic temperature of the main flow. Eckert mentioned that heat flux q_{local} in a boundary layer in a compressible flow should be defined as:

$$q_{\mathrm{local}} = h_{\mathrm{local}}[T_{\mathrm{r,local}} - T_{\mathrm{solid,local}}] \tag{82}$$

where h_{local} is the local heat transfer coefficient between a compressible flow and a solid wall. In the case where $q_{\mathrm{local}} = 0$, the local wall temperature $T_{\mathrm{solid,local}}$ equals the recovery temperature $T_{\mathrm{r,local}}$ and is called the "adiabatic wall temperature."

Here, Eckert's theory is extended to the recovery temperature T_{r} on a heating and cooling wall. In Eq. (80), the first term expressed by the static temperature T_{main} represents the internal energy that a local boundary layer originally has, and the second term expressed by the dynamic temperature $T_{\mathrm{dyn,main}} r_{\mathrm{local}}$ represents the net dynamic energy that is used to increase the temperature in a local boundary layer. When a local boundary layer is heated or cooled, the first term is affected, but the second term remains constant. The first term should be replaced by the appropriate form suitable for the heating or cooling of a boundary layer. Heating or cooling affects only a thermal boundary layer; therefore, the local total temperature $T_{\mathrm{tot,bound,x}}$ at the location x in the flow direction is defined as follows:

$$T_{\mathrm{tot,bound,x}} = T_{\mathrm{tot,in}} + \int_0^x \frac{q_{\mathrm{x}}}{\rho_{\mathrm{x}} C_{\mathrm{P}} \delta_{\mathrm{x}} u_{\mathrm{ave,x}}} dx \tag{83}$$

$$T_{\mathrm{bound,x}} = T_{\mathrm{tot,bound,x}} - T_{\mathrm{dyn,main,x}} \tag{84}$$

$$\rho_{\mathrm{x}} = \frac{P_{\mathrm{x}}}{R T_{\mathrm{bound,x}}} \tag{85}$$

http://dx.doi.org/10.5772/65931

$$\delta_x = \begin{cases} 5\left[\frac{\nu}{u_{\text{main},x}\Pr}\right]^{0.5} x^{0.5} & \text{for a laminar boundary layer} \\ 0.37\left[\frac{\nu}{u_{\text{main},x}\Pr}\right]^{0.2} x^{0.8} & \text{for a turbulent boundary layer} \end{cases} \tag{86}$$

$$u_x = \begin{cases} 0.5u_{\text{main},x} & \text{for a laminar boundary layer} \\ 0.8u_{\text{main},x} & \text{for a turbulent boundary layer} \end{cases}, \tag{87}$$

where $T_{\text{bound},x}$ and ρ_x are the static temperature and density, respectively, in a heated or cooled boundary layer, and δ_x and u_x are thermal boundary layer thickness and average velocity, respectively. Here, evaluations of δ_x and u_x are used for a plane, but more appropriate expression for a particular target flow field can be used. Finally, Eq. (80) is replaced by the following equation for the local recovery temperature of a heated or cooled boundary layer.

$$T_{\text{r,local}} = T_{\text{bound},x} + T_{\text{dyn,main},x} r_{\text{local}} = T_{\text{tot,bound},x} - T_{\text{dyn,main},x}[1 - r_{\text{local}}] \tag{88}$$

5. Mach-number distribution on solid walls with various shapes

As described in Section 4 4, the local Mach number M_{local} is constant on a plane but varies with the location on a single airfoil or an airfoil in a cascade. For a single airfoil, when the inlet Mach number M_{in}, the Reynolds number $\text{Re}_{\text{airfoil}}$ with a representative length of the airfoil chord L_C, and the angle of attack α are fixed, as

$$M_{\text{in}} = \frac{u_{\text{in}}}{a_{\text{in}}} \tag{89}$$

$$\text{Re}_{\text{airfoil}} = \frac{u_{\text{in}} L_C}{\nu_{\text{in}}}, \tag{90}$$

the distribution of the local pressure coefficient S_{local} or η_{local} is uniquely obtained. In cases of a cascade of airfoils, when the stagger angle β and the solidity σ are fixed (see **Figure 8**), the distribution of the local pressure coefficient S_{local} or η_{local} is obtained. Fortunately, many

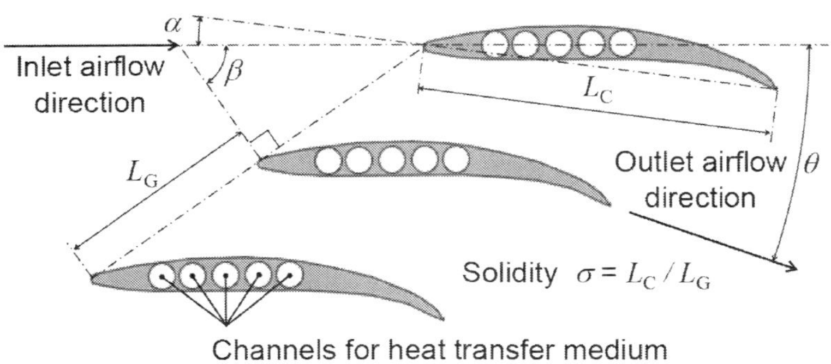

Figure 8. Flow field through a cascade of airfoils, where θ is the turning angle.

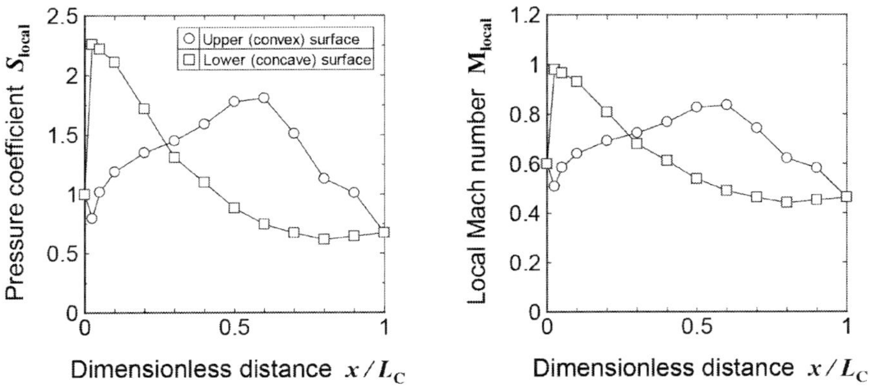

Figure 9. Local Mach-number distributions assumed from pressure-coefficient distribution.

experimental results of S_{local} or η_{local} have been reported for single airfoils and cascades of airfoils. The distributions of M_{local} are calculated using Eqs. (75), (76), and (78).

Ito et al. obtained distributions of M_{local} around an airfoil in a cascade of NACA65-($12A_2I_{8b}$)10 airfoils, as shown in the right frame of **Figure 9**, from S_{local}, which is shown in the left frame of **Figure 9** [10].

6. Air-temperature distribution in boundary layers on solid walls

Nishiyama described in his book [11] that a developing boundary layer transforms from a laminar boundary layer to a turbulent boundary layer at $Re_x \cong 10^4$ in regions with adverse

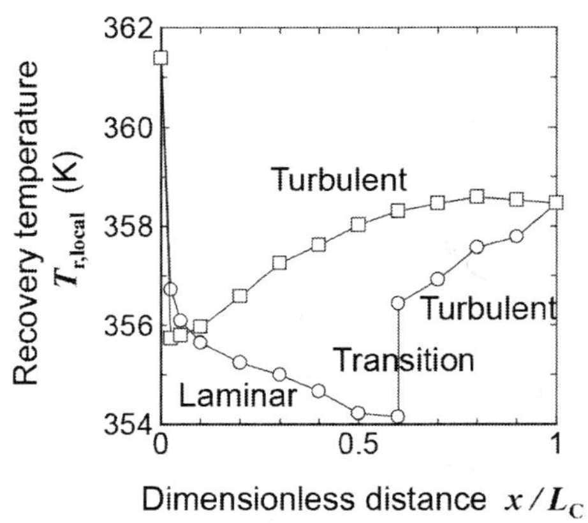

Figure 10. Recovery-temperature distribution assumed according to the pressure coefficient and local Mach number distributions in **Figure 9**.

pressure gradients, but a developing boundary layer transforms at $Re_x≅10^8$ in regions with favorable pressure gradients. This means that a developing boundary layer transforms across the minimum pressure point, that is, the maximum of the pressure coefficient S_{local} or η_{local} on the airfoil surface in cases of $Re_{airfoil}≅10^6$. According to the left graph of **Figure 9**, a developing boundary layer may transform at $x/L_C≅0.025$ on the lower concave surface and at $x/L_C≅0.6$ on the upper convex surface. Then, the local recovery temperature $T_{r,local}$ is assumed by using Eqs. (81) and (88) (see **Figure 10**). This $T_{r,local}$ can be used for the evaluation of the local heat flux q_{local} using Eq. (82) if an adequate heat-transfer coefficient h_{local} is employed.

7. Heat transfer through practical heat exchanger with complex shape

Ito et al. evaluated the rate of heat transfer from a hot compressible airflow to a cold supercritical-fluid flow through an airfoil heat exchanger, as shown in **Figure 8** [10]. Heat is transferred from the hot compressible airflow to the outer surface of the airfoil heat exchanger and is conducted from the outer surface to the five inner surfaces in the airfoil heat exchanger. Then, heat is transferred from the five inner surfaces of the airfoil heat exchanger to the cold supercritical-fluid flow inside the five tubes.

First, Ito et al. conducted wind-tunnel experiments. They installed n thermocouples into the airfoil heat exchanger and experimentally measured the temperature at n points inside the exchanger. Simultaneously, the air temperature and the air Mach number at the inlet, the supercritical-fluid temperature and pressure at the inlet, and the supercritical-fluid temperature at the outlet were experimentally measured.

Second, they assumed n heat-transfer coefficients $h_{air,1}$ to $h_{air,n}$ for the n parts of the air-contacted outer surface of the airfoil heat exchanger, as well as one heat-transfer coefficient h_{sfc} for the supercritical-fluid-contacted five inner surfaces of the airfoil heat exchanger.

Third, they performed an inverse heat-conduction analysis. The boundary conditions were set according to the experimental results for the distribution of the recovery temperature using the methods described in Sections 4.6, as well as the inlet supercritical-fluid temperature and pressure. Using these boundary conditions, heat-conduction calculations for the airfoil heat exchanger were conducted, and the temperatures at the n points in the airfoil heat exchanger and the outlet supercritical temperature were numerically obtained.

Finally, the $n+1$ numerically obtained temperatures were compared with $n+1$ experimentally obtained temperatures. If the temperatures were equal, the assumed $h_{air,1}$ to $h_{air,n}$ and h_{scf} were true. Otherwise, the assumed $h_{air,1}$ to $h_{air,n}$ and h_{scf} were corrected, and the inverse heat-conduction analysis was repeated.

Using these procedures, Ito et al. obtained an air Nusselt number correlation Nu_{air} for a cascade of NACA65-($12A_2I_{8b}$)10 airfoils, as shown in **Figure 8** [12].

$$Nu_{air} = 4.9410^{-3} Re_{airfoil} M_{in}^{1.44} \tag{91}$$

They also obtained a supercritical-fluid Nusselt number correlation Nu_{scf} for the tube flow given by Eq. (33).

Moreover, the heat-transfer rate Q_{entire} of an airfoil heat exchanger is estimated as follows:

$$Q_{entire} = \psi \kappa A_{scf} \Delta T_{lm,entire}, \tag{92}$$

where ψ is a correction factor for the airfoil heat exchanger, and ψ is the ratio of the actual heat-transfer rate to the heat-transfer rate of the ideal counter-flow heat exchanger without thermal resistance.

$$\psi = \frac{0.1236[0.02093|\xi| + 1]}{\phi_{scf} - \exp[-0.5\min\{1, \varepsilon_{SA}\}]} + 1 \tag{93}$$

Here, ξ is an incidence of air at the inlet. The incidence is a flow-direction angle from the airfoil camber (center) line at its leading edge, corresponding to an angle of attack of $\alpha = 9.47°$ for the cascade in **Figure 8**. ϕ_{scf} and ϕ_{air} indicate the temperature effectiveness, as follows:

$$\phi_{scf} = \frac{T_{scf,out} - T_{scf,in}}{T_{air,in} - T_{scf,in}} \tag{94}$$

$$\phi_{air} = \frac{T_{air,in} - T_{air,out}}{T_{air,in} - T_{scf,in}} \tag{95}$$

Here, ϕ_{scf} and ϕ_{air} are positive for an air-cooled system and negative for an air-heated system. ε_{SA} is the ratio of the heat-capacity rates.

$$\varepsilon_{SA} = \frac{m_{scf} C_{P,scf}}{m_{air} C_{P,air}} \tag{96}$$

Here, m_{scf} and m_{air} are the mass flow rates of a supercritical-fluid and air, respectively, for an airfoil heat exchanger, and $C_{P,scf}$ and $C_{P,air}$ are the specific heats of a supercritical-fluid and air, respectively. κ is the overall heat-transfer coefficient for an ideal counter-flow heat exchanger without thermal resistance.

$$\kappa = \frac{1}{\frac{1}{h_{scf}} + \frac{1}{h_{air}} \frac{A_{scf}}{A_{air}}} \tag{97}$$

Here, A_{scf} and A_{air} are areas of supercritical-fluid-contact and air-contact surfaces, respectively, for an airfoil heat exchanger. $\Delta T_{lm,entire}$ is the logarithmic mean temperature difference:

$$\Delta T_{lm,entire} = \Phi[T_{air,in} - T_{scf,in}] \tag{98}$$

Φ is the ratio of the logarithmic mean temperature difference to the temperature difference between the inlet air temperature and the supercritical-fluid temperature.

$$\Phi = 1 \qquad \text{for}\varepsilon_{SA} = 1$$
$$\Phi = \frac{|\phi_{scf}| - |\phi_{air}|}{\ln\left[\frac{\phi_{air}}{\phi_{scf}}\right]} \qquad \text{for}\varepsilon_{SA} \neq 1 \tag{99}$$

The actual heat-exchange rate is estimated as $Q_{entire}[\text{numberofairfoils}]$.

For example, Ito et al. performed cycle calculations for an intercooled and recuperated jet engine employing several pairs of airfoil heat exchangers whose heat-transfer performance is evaluated by Eqs. (91)–(99) [13].

These examples can be used for a cascade of airfoil heat exchangers; therefore, the air Nusselt number correlation in Eq. (91) or thermal resistance in Eq. (93) might be further modified in the near future according to the progress of research, as knowledge in this field is still developing.

8. Conclusion

The Nusselt number between supercritical fluid flows and solid walls can be estimated by appropriate conventional correlations using the Reynolds and Prandtl numbers if sufficiently accurate physical properties are used for each local point through the region of supercritical fluid flows. Thus, a numerical integration of local heat flow rate is required when you calculate the entire heat flow rate in a heat exchanger between supercritical fluid flows and solid walls.

The recovery temperature should be considered for the estimation of heat transfer between compressible flows and solid walls. For compressible flows on adiabatic airfoil surfaces, the local recovery temperature varies by each point on the airfoil surface, owing to the accelerating and decelerating effects of the main flow outside of the boundary layer on the airfoil surface. In addition, for compressible flows on cooling and heating airfoil surfaces, the local total temperature on airfoil surfaces in the boundary layer also varies at each point because of cooling and heating effects. The accelerating and decelerating effects can be estimated from the local Mach number distribution on the airfoil shape. The cooling and heating effects can also be estimated when a numerical integration of elapsed variation of the local total temperature along the boundary layer from the leading edge if the detailed solid temperature distribution on the airfoil surface is known. To obtain the detailed solid temperature distribution on the airfoil surface, detailed experimental measurements or an accurate CFD analysis may be required.

To estimate conjugate heat transfer through a practical heat exchanger with a complex shape, one solution is a combination of experimental results in wind tunnel tests and an inverse heat conduction method. The other solution is CFD analysis validated by experimental results in wind tunnel tests. Empirical correlations are very limited for conjugate heat transfer through a practical heat exchanger with complex shape because knowledge in this field is still developing.

Author details

Yu Ito

Address all correspondence to: itoyu110@00.alumni.u-tokyo.ac.jp

Tokyo Institute of Technology, Yokohama, Japan

References

[1] Stanley HE. Introduction to Phase Transitions and Critical Phenomena. Oxford, UK University Press; 1971.

[2] Span R, Wagner W. A New Equation of State for Carbon Dioxide Covering the Fluid Region from the Triple-Point Temperature to 1100K at Pressures up to 800MPa. Journal of Physical and Chemical Reference Data. 1996;**25**.

[3] Vesovic V, Wakeham WA, Olchowy GA, Sengers JV, Watson JTR, Millat J. The Transport Properties of Carbon Dioxide. Journal of Physical and Chemical Reference Data. 1990;**19**.

[4] Wagner W, Pruß A. The IAPWS Formulation 1995 for the Thermodynamic Properties of Ordinary Water Substance for General and Scientific Use. Journal of Physical and Chemical Reference Data 2002;**31**.

[5] Huber ML, Perkins RA, Laesecke A, Friend DG, Sengers JV, Assael MJ, Metaxa IN, Vogel E, Mareš R, Miyagawa K. New International Formulation for the Viscosity of H_2O. Journal of Physical and Chemical Reference Data 2009;**38**.

[6] Huber ML, Perkins RA, Friend DG, Sengers JV, Assael MJ, Metaxa IN, Miyagawa K, Hellmann R, Vogel E. New International Formulation for the Thermal Conductivity of H_2O. Journal of Physical and Chemical Reference Data . 2012;**41**.

[7] Dittus FW, Boelter LMK. Heat Transfer in Automobile Radiators of the Tubular Type. The University of California Publications in Engineering. 1930;**2**.

[8] Liao SM, Zhao TS: Measurements of Heat Transfer Coefficients from Supercritical Carbon Dioxide Flowing in Horizontal Mini/Micro Channels. Journal of Heat Transfer. 2002;**124**.

[9] Ito Y, Inokura N, Nagasaki T. Conjugate Heat Transfer in Air-to-Refrigerant Airfoil Heat Exchangers. Journal of Heat Transfer. 2014;**136**.

[10] Eckert ERG. Survey of boundary Layer Heat Transfer at High Velocities and High Temperature. Wright Air Development Center (WADC) TR 59–624, 1960.

[11] Nishiyama T, Yokugata Nagare Gaku [translated as "Aerodynamics of Airfoil"]. p. 23, *Nikkan Kogyo Shimbun Ltd.* (*Business & Technology Daily News*), Tokyo, 1998, in Japanese.

[12] Ito Y, Goto T, Nagasaki T. Effect of Airflow on Heat Transfer of Air-to-Refrigerant Airfoil Heat Exchanger. AIAA-2015-1193, 2015.

[13] Ito Y, Inokura N, Nagasaki T. Intercooled and Recuperated Jet Engine Using Airfoil Heat Exchangers. ISABE2015-20100, 2015.

Chapter 7

Heat Exchangers in the Aviation Engineering

Antonio Carozza

Additional information is available at the end of the chapter

http://dx.doi.org/10.5772/67486

Abstract

Heat exchangers are crucial in thermal science and engineering because of their essential role across the landscape of technology, from geothermal and fossil power generation to refrigeration, desalination, and air conditioning. In the aviation engineering, they have a fundamental role especially in reducing the temperatures of the fuel and thus increasing the efficiency of the aircraft engines. The literature on aviation heat exchangers is voluminous and continues to be updated today. Two main aspects of this class of flow systems are widely investigated: fluid flow and heat transfer performances, and criteria for evaluating those performances. In addition, the need of a smart and light equipment to be used inside a transport system is ever and ever felt. This requires a particular attention in the selection of components, for example in the engine zone, not only to reduce the weight but also to improve the whole heat transfer efficiency. With this aim, engineers focus their attention on new materials, for example porous materials, that recently have attracted researchers. The design process may be considered the heart of engineering. In this chapter, we will explore methods for the design or the choice of heat exchangers and list some practical case studies.

Keywords: pressure drop, heat transfer, heat exchanger, aircraft components cooling

1. Introduction

Heat exchangers are a fundamental tool in the thermal engineering fields, such as refrigeration, power systems cooling, electronics cooling, and air conditioning. Enhanced heat transfer (EHT) techniques provide:

- Reduction in thermal resistance (1/hA) of a conventional design with or without increase of surface area (as obtained from extended/fin surfaces)
- Passive enhancement—most commonly used method
- Active enhancement—direct input of external power

INTECH
open science | open minds

© 2017 The Author(s). Licensee InTech. This chapter is distributed under the terms of the Creative Commons Attribution License (http://creativecommons.org/licenses/by/3.0), which permits unrestricted use, distribution, and reproduction in any medium, provided the original work is properly cited.

- Compound enhancement—use of two or more methods
- Mode of heat transfer and flow regime
- Single or two phase flow, free or forced convection, laminar, or turbulent flow
- Type of application (two-fluid HE vs. single fluid HE)

The development of new kinds of heat exchangers is ever in progress, both for seeking to reduce the volume of the heat exchanger and to enhance the performances in terms of pressure drop or heat transfer capacity. Clearly, studies in these fields also consider the methods of design, the numerical tools to investigate the performances of a heat exchanger, and also the times that such a tool requires for a reasonable design. Rohsenow et al. [1], Kays and London [2], and Bejan and Lorente and Bejan et al. [3, 4] have largely investigated these issues. Yang et al. [5], You et al. [6], Caputo et al. [7] have given valid examples of optimization and design techniques related to heat exchangers. The process of design of aviation heat exchangers occurs with an increase in the power stress and servicing energy systems that must be cooled. At moderate flight speeds, the systems of aviation engines, lubricants, and different equipment and optional energy systems can be cooled with heat exchangers using free air or other carriers. The design should provide both the satisfaction of such constructional requirements as the compensation of different thermal extensions of heat exchange surfaces and the body, the heat exchanger compactness and the possibility of assembling heat exchangers with the use of the existing equipment in a reliable manner. These requirements are often contradictory. That is why, when designing heat exchangers, it is very important to determine the optimality criteria in each particular case. Heat exchangers may be classified according to the following main criteria:

- Recuperators and regenerators
- Transfer processes: direct contact and indirect contact
- Geometry of construction: tubes, plates, and extended surfaces
- Heat transfer mechanisms: single phase and two phase
- Flow arrangements: parallel, counter, and cross-flows

First of all, heat exchangers can be distinguished considering either an intermediate storage or a direct transfer of heat (see also **Figure 1**). In a regenerator, the heat coming from the primary medium is first stored in a medium playing as a reservoir and then regenerated from that mass by the secondary medium. The reservoir material can be the one of the ducts or a porous medium through which the primary and the secondary flows are driven.

In a recuperator both the media are separated by a wall through which heat is transferred directly. Further an intermediate medium can carry heat from the primary medium in a first heat exchanger to the secondary medium in a second heat exchanger.

The regenerator needs unfortunately an intermediate storage material that is a good heat conductor for the storage function. This generates high heat conduction levels in the flow direction producing a substantial loss of effectiveness ($\ll$90%). In a recuperator, instead, the only fundamental loss is the heat conduction through the wall in the flow direction, which however can be

http://dx.doi.org/10.5772/67486

reduced to less than a per mille by using materials with low thermal conductivity such as plastics. If we want to achieve the required effectiveness, only a recuperator can be used.

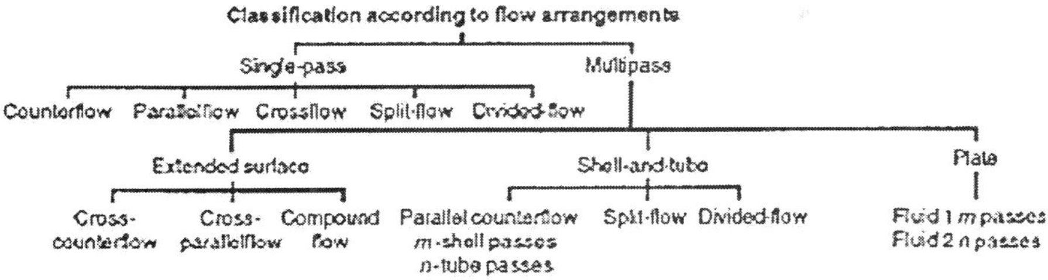

Figure 1. HE classification based on flow arrangements [2].

When classified according to the transfer process involved, heat exchangers can be a direct-contact type or an indirect-contact type. The most common type employed is the indirect-contact heat exchanger. In direct-contact exchangers, heat transfer between fluids occurs through direct interaction, ideally without mixing or leakage. The fluids come into direct contact, exchange heat, and are then separated. Advantages of these include a low cost and a lack of fouling (absence of transfer surface), the major drawback being the fact that applications are limited to situations in which direct contact of fluid streams is viable. They are particularly useful in applications involving mass transfer in addition to heat transfer, obtained through fluid phase change; heat transfer involving only sensible heat is rare for this type of exchanger. Due to the increased enthalpy, latent heat transfer is responsible for the greater portion of energy transferred in this process.

The selection of the flow arrangement influences the overall performance of the heat exchanger and is influenced by available pressure drops, permissible velocities, and thermal stresses, flow paths, required temperature levels, amongst others. The following diagram establishes a classification of heat exchangers based on available fluid stream flow types.

In the single-pass setup, fluids enter the exchanger and come into thermal contact once before proceeding to exit the device. Amongst single-pass exchangers, the counter-flow configuration is the most efficient, producing the highest temperature change in each fluid. In these, the fluids flow parallel to each other, but in opposing directions. The parallel flow type, in which fluid streams enter together at the end, is the least efficient of the single-pass devices. A more unique flow arrangement is the cross-flow, fluids flowing in normal directions to each other, most common in extended transfer surfaces, leading to two-dimensional temperature variations.

As for the multipass arrangement, the fluids essentially transit through the heat exchanger on more than one occasion, using two or more passages for each fluid, in order to achieve this. Multipass design is particularly useful in situations requiring extreme exchanger length or low fluid velocities. A great advantage of this type of exchangers is the increase in overall efficiency that results from increasing the effectiveness in each pass, resulting the greater thermal transfer load. As would be expected, at some stage of the fluid trajectory, reversal must take place. This task is typically accomplished by the introduction of U-bends in the fluid passages, which dismisses the requirement of additional external power sources. The multipass configuration can be further classified in order of construction type: extended surface (in which fluid cross-flow is the typical arrangement), shell and tube (common domain for U-bend employment), and plate exchangers [8–10].

To save the consumption of fuel and for efficient cooling, one needs to keep the heat exchangers clean for smooth functioning. An aircraft for the transport of passengers has often a pleasant cabin environment for a comfortable journey. Heat exchangers are commonly used to cool hydraulics, rammed air, auxiliary power units, gearboxes, and many other components that consist of an aircraft. Although temperature is the main feature associated with liquid cooling, when heat exchanger services are used at high altitudes air density and pressure are additional features considered. In order to guarantee a sufficient airflow, heat exchanger's fan must be carefully selected based on the ambient pressure. At high altitudes, the density of air is drastically lower, so it takes more airflow to remove the same amount of heat. Liquid cooling can provide notably better performance than air cooling, along with a quieter behavior and not vulnerable to altitude. They also reduce weight and power consumption avoiding the need for large fans or the need for wide spacing for placing components. Heat exchangers, liquid cooled chassis, and cold plates are used to provide thermal solutions to cool aircraft fluids and electronic equipment. Also, air is significantly colder than at sea level on high altitudes. Novel compact heat exchanger (CHE) solutions are needed in aerospace environmental control, avionics, and engine oil cooling systems, see **Figure 2**.

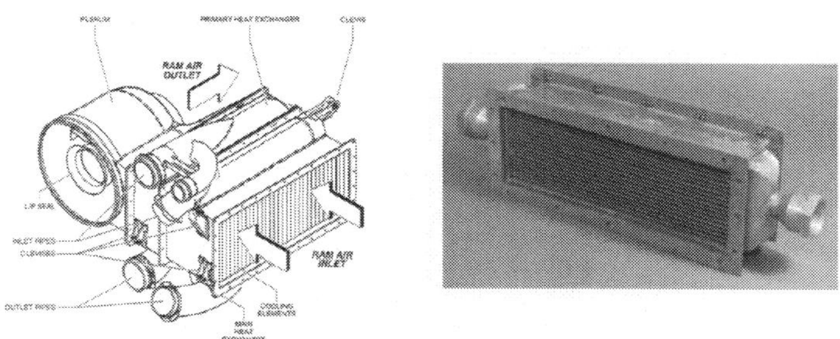

Figure 2. Heat exchangers in a typical aircraft air conditioning (ACS) pack—a typical single-pass, cross-flow, plate fin heat exchanger.

Heat exchangers are compact generally when the heat transfer area per unit volume is >700 m^2/m^3. Nowadays, the compact heat exchangers are increasingly used in a large number of industries. The application determines the material of construction, fabrication, and development of the compact heat exchangers. Inside the compact heat exchangers, fluids interact with a much larger surface area which provides higher heat transfer rates and large effectiveness. These heat exchangers are particularly suitable for the aerospace applications due to their less weight, greater compactness, and higher performances due to the improved heat transfer surfaces. In aerospace industries, furthermore, attention is paid on size and weight without compromising on performance aspects, and these compact heat exchangers are principally utilized. A compact heat exchanger also includes thin plates and fins which are stacked together and are normally brazed or welded. The aircraft heat exchangers during their operation also can meet adverse ambient conditions [11].

The aircraft heat exchangers experience arduous and extreme working conditions during their operation. Hence, the mechanical integrity and endurance life of heat exchanger need to be estimated before leading to flight clearance. The compact heat exchanger (CHE) is characterized by a small volume and a high rate of energy exchange between two fluid streams by employing intricate flow passages. Thermohydraulic performances of compact heat exchangers are strongly depended upon the prediction of performance of various types of heat transfer surfaces, such as offset strip fins, wavy fins, rectangular fins, triangular fins, triangular, and rectangular perforated fins in terms of colburn "*j*" and fanning friction "*f*" factors. Earlier, these data could be generated only through a dedicated experimental test rigs.

Now, the numerical methods play a major role for analysis of compact plate-fin heat exchangers, which are cost effective and fast. The aerospace applications—microchannel heat exchangers have the following characteristics:

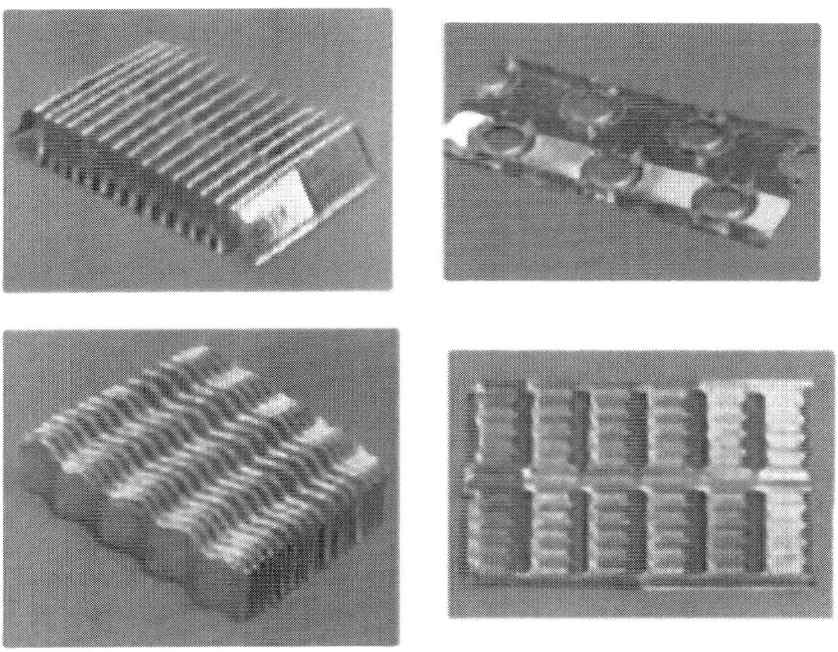

Figure 3. Examples of microchannels.

- Flow channel geometry/shape is important for high performance microchannel HEs (see **Figure 3**).
- Additive manufacturing or improvements in conventional fabrication methods are needed to resolve current challenges.
- Additional benefits can be obtained by incorporating compound enhancement methods.

The features of the most commonly used heat exchangers in aviation are listed in **Table 1**. It shows that the materials most often used are aluminum, copper, and carbon steel, while the typical sizes range between 100 mm and 132 cm.

Materials of construction
Aluminum
Carbon steel
Copper
Cupronickel
Nickel alloys
Stainless steel
Fluids commonly worked with air
Coolants
Petroleum products
Refrigerants
Water
Heat transfer capacity (typical)
200 W to 300 KW
Typical unit size range
Depths from 0.75 in. (19 mm) to 24 in. (61 cm)
Heights from 1 in. (25 mm) to 52 in. (132 cm)
Widths from 4 in. (100 mm) to 52 in. (132 cm)

Table 1. Materials and commonly used fluids and sizes in aviation heat exchangers.

2. Design

Based on the problem specifications, the heat exchanger construction type, flow arrangement, surface or core geometry, and materials must be selected. The most common problems in heat exchanger design are rating and sizing. This chapter discusses the basic design methods for two fluid direct transfer heat exchangers. The rating problem is concerned with the determination of the heat transfer rate and the fluid outlet temperatures for prescribed fluid flow rates, inlet temperatures, and allowable pressure drop for an existing heat exchanger; hence, the heat transfer surface area and the flow passage dimensions are available. The sizing problem, on the other hand, involves determination of the dimensions of the heat exchanger, that is, selecting an appropriate heat exchanger type and determining the size to meet the requirements of specified hot and cold fluid inlet and outlet temperatures, flow rates, and pressure drops. Problem definition and design process passes through the following parameters choice:

- Number, materials, and flow arrangement of heat exchangers
- Size and geometrical properties of the surfaces
- Operating conditions
- Thermophysical properties of fluids and materials
- Main design variables: pressure drop and heat transfer
- Complementary design variables: vibrations and thermal stresses
- Evaluation criteria of economical and engineering kind toward a compromise solution.

See the heat exchanger design methodology in Ref. [9] and summarized in next sections (see **Figure 4**).

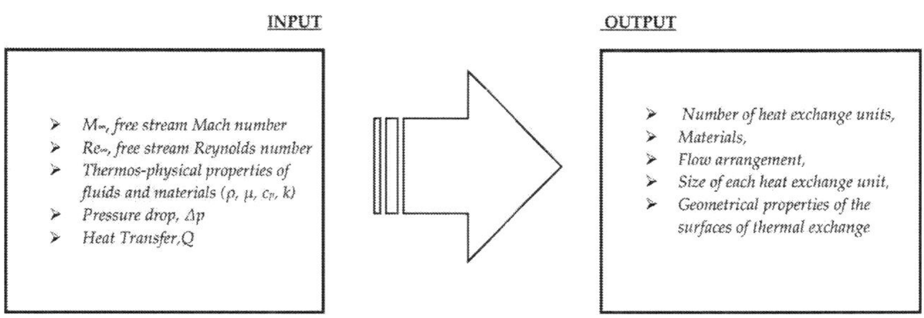

Figure 4. Design loop of a heat exchanger.

2.1. Problem definition

Generally the heat transfer rate in a heat exchanger can be calculated by

$$Q = \Delta T_{lm} U \beta V \tag{1}$$

Therefore, improvements of heat transfer can be achieved by increasing exchanger volume V, area density β of the exchanger, logarithmic mean temperature difference ΔT_{lm}, or overall heat transfer coefficient U, including the heat transfer coefficients and the conductivity of the wall. The convective heat transfer coefficient of gases usually is one or two orders of magnitude lower than that of liquids. For this reason, a high heat transfer area is necessary for realizing a high heat transfer rate, especially if one or more fluids are gaseous. This means the surface must be compact. It is defined that a heat exchanger is compact, if it incorporates at least one compact surface. On the other hand, heat exchangers with densities of 6600 m^2/m^3 are also used. The logarithmic mean temperature is calculated by the formula:

$$\Delta T_{lm} = \frac{(\Delta T_H - \Delta T_C)}{ln\left(\frac{\Delta T_H}{\Delta T_C}\right)} \tag{2}$$

Log-mean temperature difference (LMTD) is a good measure of the effectiveness of similar heat exchangers of different designs. Often, LMTD (counter flow) > LMTD (parallel flow). When there is insufficient information to calculate the log-mean temperature difference (LMTD), the so-called number of transfer units (NTU) method is used to calculate the rate of heat transfer in heat exchangers (especially countercurrent exchangers). In heat exchanger analysis, if the fluid inlet and outlet temperatures are specified or can be determined by simple energy balance, the LMTD method can be used, but when these temperatures are not available. The NTU or the effectiveness method is used. The maximum heat transfer rate, Q_{max}, is evaluated by

$$Q_{max} = (mc)_{min}(T_{h,i} - T_{c,i}) \tag{3}$$

The heat transfer rate of the heat exchanger, Q, is calculated by

$$Q_c = m_c c_c (T_{c,o} - T_{c,i}) \tag{4}$$

The effectiveness (NTU method) ε is calculated by

$$\varepsilon = \frac{Q}{Q_{max}} \tag{5}$$

For more details, the reader can refer to the works [12, 13]. Heat flux is calculated by

$$q = \frac{Q}{A} = \frac{m_c c_c (T_{c,o} - T_{c,i})}{nL_c W_c} \tag{6}$$

or

$$q = U\Delta T_{lm} = \frac{\Delta T_{lm}}{\sum R} \tag{7}$$

the overall thermal resistance $\sum R$ is determined by

$$\sum R = R_{cond} + R_{conv} \tag{8}$$

where $\dot{m}$ is the mass flow rate (subscripts h and c stand for the hot side and cold side, respectively), n is the number of channels, c is the specific heat, $T_{h,i}$, $T_{h,o}$, $T_{c,i}$, and $T_{c,o}$ are inlet and outlet temperatures of the hot and cold side, respectively, q is the heat flux, A is the heat transfer area, k is the overall heat transfer coefficient, $R_{cond} = \delta/\lambda$ is the conductive thermal resistance, $R_{conv} = 1/h_h + 1/h_c$ is the convective thermal resistance, h_h and h_c are the convective heat transfer coefficients of the hot side and the cold side, respectively, δ is the thickness, and λ is the thermal conductivity [14].

Another parameter that is really important during the design process of a cooling system is the total pressure drop, sum of a cold side pressure drop, and hot side pressure drop,

$$\Delta p_{tot} = \Delta p_c + \Delta p_h \tag{9}$$

Manufactures of heat exchangers have then the possibility to choose a wide variety of materials, such as aluminum, stainless steel, copper, and cupronickel, chosen based upon heat transfer or environmental requirements

The Reynolds number Re is calculated by

$$Re = \frac{\rho w D_h}{\mu} = \frac{2\dot{m}}{\mu(W_c + D_c)} \tag{10}$$

pressure drop due to friction is determined by [14, 15]:

$$\Delta p = 2f\rho u^2 \frac{L}{D_h} = 2fRe\frac{L}{D_h{}^2}w\mu \tag{11}$$

where $D_h = 4A_c/P$ is the hydraulic diameter, u is the flow velocity, μ is the dynamic viscosity, ρ is the density, A_c is the cross-sectional area, P is the wetted perimeter, L is the channel length, and f is the Fanning friction factor [15].

The performance index, ξ, is determined by [16–21]

$$\frac{Q_c}{\Delta p_t} = \frac{m_c c_c(T_{c,o} - T_{c,i})}{\Delta p_h + \Delta p_c} \tag{12}$$

In aviation, the cold fluid is often represented by air, whereas the hot fluid is for example the oil that must reduce the engine temperatures. The real parameter of choice is the group hA based on the following simplified equation:

$$Q = hA(T_w - T_f) \tag{13}$$

where T_w and T_f are the hot wall temperature and the cooling air temperature, respectively. First of all, an aerospace engineer tries to reduce the size of a heat exchanger for a fixed heat duty. To contain the HE size, the surface geometry is designed in order to increase the heat transfer area, using the so-called compact heat exchangers, that can be of plate fin or tube fine typology, see Ref. [22]. A particular surface geometry affects both the heat transfer coefficient h and also the pressure drop Δp. These parameters are correlated with the flow regime of each fluid and so they depend on the value of Reynolds number Re. A useful correlation in this sense can be given by the colburn factor that can be calculated by

$$j = \frac{h}{\rho w c_p} Pr^{2/3} \tag{14}$$

where c_p is the air specific heat, h is the convective heat transfer coefficient, w is the air flow velocity, ρ is the density, and $Pr = v/\alpha$ is the Prandtl number in the flow conditions under investigation, v is the kinematic viscosity, and α is the thermal diffusivity.

2.2. Computational methodologies

The numerical methods that can be used to investigate the deeper behavior of a heat exchanger are often based on the Navier-Stokes solution rather than on a balance of heat fluxes and mass flow rates.

2.2.1. Governing equations

The domain under investigation must be discretized my means of a preprocessor—grid generator and on it the following equations are to be manipulated numerically in order to have the fluid and thermal fields of the case, as follows:

Mass

$$\frac{\partial \rho}{\partial t} + \frac{\partial}{\partial x_j}[\rho u_j] = 0 \tag{15}$$

Momentum

$$\frac{\partial}{\partial t}(\rho u_i) + \frac{\partial}{\partial x_j}[\rho u_i u_j + p\delta_{ij} - \tau_{ji}] + S_{F,i} = 0, \ i = 1,2,3 \tag{16}$$

Energy

$$\frac{\partial}{\partial t}(\rho e) + \frac{\partial}{\partial x_j}[\rho u_j e + u_j p - u_i \tau_{ij}] + S_E = 0 \tag{17}$$

Based on the typology of flow regime, other equations can be solved to take into account the turbulent fluctuations. It must be reminded that the flow regime is strictly related to the heat transfer that can be achieved in the heat exchanger. Often other heat sources can play a fundamental role in the solution of the aforementioned equations. In this case, other equations have to be added to the basic ones increasing the times to reach the solution, so before proceeding to solve the equations governing the fluid-dynamics inside and out of the cooling ducts, the right evaluations in terms of time and number of simulations must be done.

2.2.2. Design approaches

Often the manufacturers offer characteristic curves of a heat exchanger on their catalogs so to make possible a preliminary selection of the right choice. Nevertheless, this is not sufficient and so the designer has to deep analysis using iterative procedures that pass through a number of numerical simulations or graphical assessments.

2.2.3. *Numerical tools*

There are a number of numerical tools capable of simulating a heat exchanger, some simplified and other with the possibility of considering more details. They are dedicated codes or numerical models inserted in more complex numerical packages. Often the design can require more software, from computer-aided design to computational fluid dynamics solvers. The most important parameters to monitor during an HE simulation are follows:

- pressure drop, Δp
- mass flow rate, m
- heat exchange, Q
- working fluids properties, ρ, μ, c_p, k
- flow directions

Heat exchangers are designed to maximize the surface area of the wall between two fluids, while minimizing resistance to fluid flow through the exchanger by means of thermal analyses, CFD, and FEA, to ensure efficient and effective optimized designs. Different commercial codes are present on the market:

- ANSYS fluent,
- ANSYS workbench,
- COMSOL multiphysics,
- StarCCM+,

They include packages able to simulate the flow fields inside the tubes and also specific numerical models capable of simulating special fluids like nanofluids or special structures like porous media [23–26].

2.3. Economic evaluation

Once the heat exchanger size and characteristics are chosen, the designer has to proceed to an economical evaluation of this choice in terms of maintenance costs and, if the case, reinstallation costs. The components inside the heat exchangers have to be free from deposits and dirt built up during flying operations. This is vital because substandard cleaning could result in a loss of pressure in the heat exchanger, which is unacceptable. Therefore, they need to be cleaned at regular intervals. In the past years, airline companies needed to hire engineers who would conduct elaborate investigations into the dirt accumulation and physical/chemical surface analysis of the aluminum plates in the center. Now, there are agents for heat exchanger services who perform it by using a scanning electron microscope to recognize the different elements of the mount up dirt. Nevertheless, maintenance and reinstallation involve often higher costs that must be considered in the economic evaluation of the HE typology to choose already in the preliminary design phase.

3. Case studies

Air cooling is accomplished by air flowing into the engine compartment through openings in front of the engine cowling. Expulsion of the hot air occurs through one or more openings in the lower portion of the engine case. The air cooling system is less effective during ground operations, takeoffs, go-arounds, and other periods of high-power, low-airspeed operation. The engine should not operate at higher than its designed temperature because of loss of power and excessive oil consumption, see a list of engine typologies in **Figure 5**. Two case studies are presented in the following subsections regarding the best possible mass flow rate ensuring the reduction of the temperature under critical flight conditions.

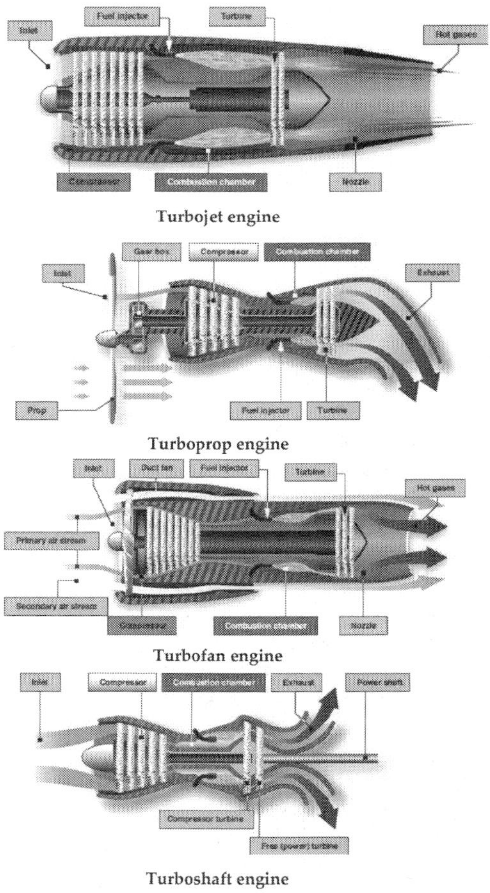

Figure 5. Different typologies of aviation engines.

3.1. Heat exchanger efficiency in a pusher engine

We verify here the air flow rate efficiency of an aircraft oil cooling system. The objective of this analysis is the computation of the mass flow rate useful for a regional new concept pusher engine aircraft under cruise conditions. A simplified geometry is used in order to adopt a porous medium numerical model. This tool allows to model the pressure losses and the heat transfers using the input parameters of a software package for a Darcy-Forchheimer porous material. The mass flow rate to be achieved in order to reduce the engine temperatures is

m, mass flow rate target = 0.25 kg/s

The following table illustrates the heat exchanger performances:

The flow condition (cruise) has the following characteristics:

- Mach, $M_\infty = 0.28$
- Reynolds, $Re_\infty = 4.5$ millions
- Ambient pressure, $p_\infty = 72{,}500$ Pa
- Ambient temperature, $T_\infty = 265$ K

The classical Darcy-Forchheimer law for porous materials reads:

$$\frac{\Delta p}{L} = \frac{\mu}{K} v + \frac{\rho}{2} C_2 v^2 = av + bv^2 \tag{18}$$

where K is the permeability of the material, μ is the related dynamic viscosity, a is the Darcy coefficient, and b is the inertial Forchheimer coefficient. The relationship between the mass flow rate $\dot{m}$ and pressure drop Δp in the cooler can be put in a similar shape, see the values shown in **Table 2**. Nevertheless, as indicated in equation Eq. (18), viscous and inertial resistances are completed as follows:

Viscous resistance

$$20.598 = \frac{\mu}{K} \Delta n \rightarrow a = 542650.18 \ 1/\mathrm{m}^2 \tag{19}$$

Inertial resistance

$$2.1355 = C_2 \frac{1}{2} \rho \Delta n \rightarrow b = 23.10 \ 1/\mathrm{m} \tag{20}$$

where Δn is the HE total length in the flow direction, ρ is the air density, and μ is the air dynamic viscosity, see papers [23–26]. By using the aforementioned values, one can simulate the external and internal flow fields with respect to the cooling ducts and find the numerical air flow rate passing through the heat exchanger. In this manner, one can establish if the aforementioned target is achieved or not and, in this last case, choose another heat exchanger.

Mass flow rate $\dot{m}$ [kg/s]	Pressure drop Δp [Pa]	Heat eject $\dot{Q}$ [kW]
0.1	220	200
0.2	356	300
0.3	520	450
0.4	720	550
0.5	860	650

Table 2. Characteristic values of the heat exchanger taken into consideration.

3.2. Compact heat exchangers choice for a light helicopter

Often the choice of the heat exchanger positioning can be crucial in a life of a helicopter. In the following, a number of simulations are presented tailored to the best cooler location for a light helicopter. The flow equations are solved by means of a finite volume code and a structured grid generator.

Three positions for the heat exchanger location have been investigated by a thermal point of view as shown in **Figure 6**. Each position is characterized by a specific value of air flow rate. In order to maximize the flow rate and taking into account also maintenance problems, the manufacturer chooses to locate the cooler near the opening after the engine because the cooler is crossed by a mass flow rate able to cool the engine and can be inspected by the maintenance team. Using the same methodology of the previous paragraph, the designer, in a simplified manner, and without knowing the constructive details of the heat exchanger, can so take preliminary decisions about the size and the positioning of the heat exchanger reducing the number of certification tests and the related cost, see also paper [25].

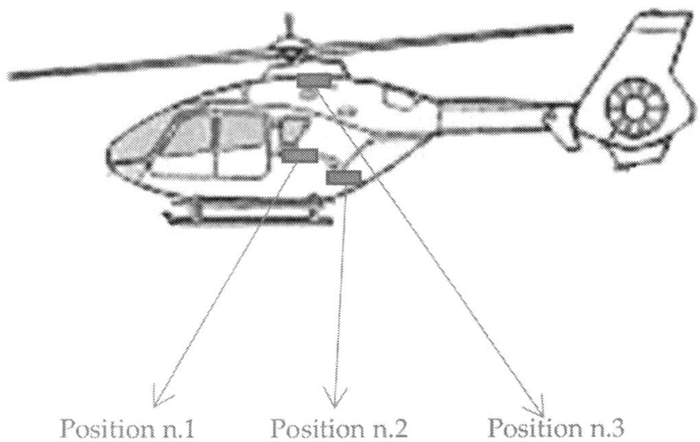

Figure 6. Light helicopter side view with heat exchanger locations.

4. Conclusion

The aim of this chapter is to serve as a simplified guide, which is useful for the designers in the preliminary choice and sizing of a heat recovery unit. It is composed of a preliminary introduction summarizing all the types of heat exchangers and illustrates the typical size of the components of an aviation cooling unit. Then the characteristic variables of a recovery unit are listed together under the flow condition parameters to be used for simulating properly the ambient flow field in which the heat exchanger has to work. Finally, two test cases are shown and described to make the reader able to have an idea of how to set such a problem. The whole chapter can be summarized in three main points, as follows:

- Choose the best design method useful to analyze the cooling performances of a heat exchanger.
- Establish the inputs and the targets to reach at the end of such a design process.
- Establish what is the better exchanger and eventually reduce the time to production of the aircraft.

Nomenclature

a	Speed of sound, m/s
A	Heat exchange surface area, m^2
A_c	Cross-sectional area, m^2
CHE	Compact heat exchanger
CFD	Computational fluid dynamics
C_2	Inertial porous coefficient
c_p	Specific heat, J/kg/K
Δn	Heat exchanger thickness, m
D_h	Hydraulic diameter, m
e	Energy per mass unit, W/kg
f	Friction factor
EHT	Enhanced heat transfer
FEA	Finite element analysis
h	Heat transfer coefficient, W/m^2K
i,j	One of the Cartesian directions (x,y,z)
K	Permeability, 1/m
k_r	Thermal conductivity ratio
k_v	Volumetric heat transfer coefficient, W/m^3K
j	Colburn factor
L	Length of channel, m
$M=u/a$	Mach number
m	Mass flow rate, kg/s
n	Number of tubes

NTU	number of transfer unit
Nu	Nusselt number
p	Pressure, Pa
$Pr = \upsilon/\alpha$	the Prandtl number
Q	Heat transfer rate, W
q	Heat flux, W/m^2
R	Thermal resistance, m^2K/W
$Re = uL/\upsilon$	Reynolds number
σ	Thermal stress, Pa
$S_{F,i}$	Momentum source along the ith Cartesian direction, Pa
S_E	Energy source, W
T	Temperature, K
T_d	Mean temperature, K
u	Fluid local velocity, m/s
U	Overall heat transfer coefficient, W/m^2K
$\dot{V}$	Volume flow rate, m^3/s
x	Cartesian coordinate, m

Greek symbols

β	Area density, m^2/m^3
δ_{ij}	Kronecker symbol
ε	Effectiveness
λ	Thermal conductivity, W/m K
μ	Dynamic viscosity, Ns/m^2
Δp	Pressure drop, Pa
ρ	Density, kg/m^3
ξ	Performance index, W/kPa
τ	Viscous stresses, Pa
ΔT	Different temperature, K
∞	Free stream

Subscripts

cond	Conductive
conv	Convective
c	Cold
h	Hot
f	Cooling air
i	Inlet
lm	Logarithmic mean

http://dx.doi.org/10.5772/67486

o Out
tot Total
w Hot wall

Author details

Antonio Carozza

Address all correspondence to: a.carozza@cira.it

Department of Industrial and Information Engineering, Aversa, Italy

References

[1] Rohsenow WM, Hartnett JP, Ganic EN. Handbook of Heat Transfer Applications. 2nd ed. New York: McGraw-Hill; 1985.

[2] Kays WM, London AL. Compact Heat Exchangers. 3th ed. New York: McGraw-Hill; 1984.

[3] Bejan A, Lorente S. Design with Constructal Theory. Hoboken: Wiley; 2008.

[4] Bejan A, Lorente S, Yilbas BS, Sahin AZ. The effect of size on efficiency: power plants and vascular designs. Int. J. Heat Mass Transfer. 2011; **54**: 1475–1481.

[5] Yang J, Oh S, Liu W. Optimization of shell-and-tube heat exchangers using a general design approach motivated by constructal theory. Int. J. Heat Mass Transfer. 2014; **77**: 1144–1154.

[6] You Y, Chen Y, Xie M, Luo X, Jiao L, Huang S. Numerical simulation and performance improvement for a small size shell-and-tube heat exchanger with trefoil-hold baffles. Appl. Therm. Eng. 2015; **89**: 220–228.

[7] Caputo AC, Pelagagge PM, Salini P. Heat exchanger design based on economic optimization. Appl. Therm. Eng. 2008; **28**: 1151–1159.

[8] Incropera FP, DeWitt DP. Fundamentals of Heat and Mass Transfer. 5th ed. New York: Wiley; 2001.

[9] Sadic Kakaç. Thermal Hydraulic Fundamentals and Design. In: Heat exchangers. Washington DC: Hemisphere; 1981. pp. 455–459.

[10] Ramesh K Shah, Sekulic PD. Fundamentals of Heat Exchanger Design. Hoboken: Wiley; 2003.

[11] Pellischek G. Compact heat exchanger technology for Aero Engine. In: Proceedings of the 10th International Symposium on Air Breathing Engines; 1–6 September 1991; Nottingham, England: AIAA; 1991, pp. 203–208.

[12] Incropera F P, DeWitt D P. Fundamentals of Heat and Mass Transfer. 3rd ed. New York: Wiley; 1990. pp. 658–660.

[13] Incropera FP, DeWitt DP, Bergman TL, Lavine AS. Fundamentals of Heat and Mass Transfer. 6th ed. New York: Wiley; 2006. pp. 686–688.

[14] Jiji LM. Heat Convection. 2nd ed. Berlin Heidelberg: Springer; Verlag; 2009.

[15] Kandlikar SG, Garimella S, Li D, Colin S, King MR. Heat Transfer and Fluid Flow in Minichannels and Microchannels. Singapore: Elsevier Pte Ltd; 2006.

[16] Hasan MI, Rageb AA, Yaghoubi M, Homayoni H. Influence of channel geometry on the performance of a counter flow microchannel heat exchanger. Int. J. Therm. Sci. 2009; **48:** 1607–1618.

[17] Ameel TA, Warrington RO, Wegeng RS, Drost MK. Miniaturization technologies applied to energy systems. Energy Convers. Manage. 1997; **38:** 969–982.

[18] Dang TT, Chang YJ, Teng JT. A study on the simulations of a trapezoidal shaped micro heat exchanger. J. Adv. Eng. 2009; **4:** 397–402.

[19] Dang TT, Teng JT, Chu JC. Effect of flow arrangement on the heat transfer behaviors of a microchannel heat exchanger. In: Proceedings of the International Multi Conference of Engineers and Computer Scientists 2010 (IMECS2010); Hongkong; **2010**; 2209–2214 (best student paper award).

[20] Dang TT, Teng JT. Numerical and experimental studies of the impact of flow arrangement on the behavior of heat transfer of a microchannel heat exchanger. IAENG Int. J. Appl. Math. 2010; **40:** 207–213.

[21] Dang TT, Teng JT. Influence of flow arrangement on the performance for an aluminium microchannel heat exchanger. IAENG Transactions on Engineering Technologies. The American Institute of Physics (AIP). 2010; **1285**: 576–590.

[22] Kakac S, Liu H. Heat Exchangers Selection, Rating and Thermal Design. 2nd ed. Florida: CRC Press; 2002.

[23] Carozza A, Mingione G, Pezzella G. Computational flow field analyses on aeronautical oil cooling systems. In: Proceedings of the 21th AIDAA Conference; September 2013; Napoli.

[24] Carozza A. An innovative approach for the numerical simulation of oil cooling systems. Adv. Aircraft Space. Sci. Int. J. 2015; **2**:169–182.

[25] Carozza A. Assessment of the performances of a heat exchanger in a light helicopter. Adv. Aircraft Space. Sci. 2015; **2**(4):469–482.

[26] Carozza A. Numerical investigation of heat exchanger performances for a two engine aircraft in pusher configuration by an U-RANS code. Appl. Therm. Eng. 2016; **99**: 1048–1056.

Chapter 8

Operation and Performance Analysis of Steam Generators in Nuclear Power Plants

Siniša Šadek and Davor Grgić

Additional information is available at the end of the chapter

http://dx.doi.org/10.5772/66962

Abstract

Steam generators are components in which heat produced in the reactor core is transferred to the secondary side, the steam supply system, of the nuclear power plant (NPP). Steam generators (SGs) have to fulfil special nuclear regulatory requirements regarding their size, selection of materials, pressure loads, impact on the NPP safety, etc. The primary-side fluid is liquid water at the high pressure, and the fluid on the secondary side is saturated water-steam mixture at the pressure twice as low. A special attention is given to preserving the boundary between the contaminated water in the primary reactor coolant system and the water-steam mixture in the secondary system. A brief overview of the SG design, its operation and the mathematical correlations used to quantify heat transfer is given in the chapter. Results of the SG transient behaviour obtained by the simulation with the best-estimate computer code RELAP5, developed for safety analyses of NPPs, are also presented. Two types of steam generators are analyzed: the inverted U-tube SG, which is commonly found in the present-day pressurized water reactors and the helical-coil SG that is part of the new-generation reactor designs.

Keywords: steam generator, nuclear power plant, reactor safety, numerical analysis, SG operation, inverted U-tube SG, helical-coil SG

1. Introduction

Steam generators (SGs) are nuclear power plant components (NPPs) in which the steam, driving the turbine, is produced. They are heat exchangers where the heat produced in the reactor core is transferred to the secondary side, the steam system, of the nuclear power plant. Their other important function is to provide the barrier between the reactor coolant, which may be contaminated with radioactive fission products, and the environment. The thin tubes with a large surface area act both as heat transfer elements and fission product

INTECH
open science | open minds

© 2017 The Author(s). Licensee InTech. This chapter is distributed under the terms of the Creative Commons Attribution License (http://creativecommons.org/licenses/by/3.0), which permits unrestricted use, distribution, and reproduction in any medium, provided the original work is properly cited.

barriers. The priority of a nuclear safety philosophy is to maintain a sufficient water level in the SGs to avoid the damage of the tubes and release of radioactive fluid from the NPP. Thus, there are two functional requirements placed on the steam generators. The first is to act as a heat sink for the reactor core to prevent any core damage. The second functional requirement is to generate the flow rate of steam from the feedwater supply at the temperature, pressure and enthalpy conditions necessary to efficiently drive the steam turbine/electric generating system.

There are few steam generator designs depending on the type of the nuclear power plant [1]. Here, it is necessary to mention that the discussion inside the chapter is focused solely on heat exchangers in the so-called pressurized water reactor (PWR) NPPs where both the primary and secondary fluids are water. There are some other NPP types with gases, salts, or liquid metals used as reactor coolants, but they are not going to be presented. There are also light water reactor power plants where water boils directly inside the reactor core, boiling water reactors (BWRs), which do not have the steam generators at all, because the steam is produced in the reactor itself. Nonetheless, PWRs are the most common type of NPPs making up to 63% of the total number worldwide, while BWRs make up 18% of the total.

The most common design is the SG with the inverted U-tube heat exchanger bundle, where steam separation equipment is located inside the top shell of the SG [2]. The primary water flows upwards through the tubes first, making a bend, and then flows downwards back into the reactor coolant system (RCS) piping. On the secondary side, roughly a quarter of water, injected through the feedwater system, evaporates, and the remaining water is recirculated back into the boiling region; therefore, that type of SG is called recirculating steam generator (RSG). The American company Babcock and Wilcox (B&G) developed the once-through steam generator (OTSG), a vertical shell counterflow straight-tube design which directly generates superheated steam as the feedwater flows through the SG in a single pass. The primary-side water enters at the steam generator at the top, flows through the generator in unbent tubes and exits at the bottom. There is also a horizontal type of the SG which has the horizontal cylindrical housing and horizontal coils. The steam is dried at the top of the housing by gravitational separation. That type of SG is used mostly in Russian types of PWR reactors called vodo vodijanoj energetičeskij reactor (VVER). The last important design, to be later analyzed in the chapter, is the vertical SG with helical tubes, which has the intention to be used in the future advanced modular and high-temperature reactor systems [3]. The helical-coil design offers compactness and increased heat transfer. At the SG outlet, the steam is superheated, which results in higher thermal efficiency.

2. Description and operation of recirculating steam generators

2.1. Design configuration and flow

A recirculating steam generator consists of a bottom head of a hemispherical shape, the central cylindrical part where the heat exchange occurs and the upper part, the steam dome. The lower head is divided into inlet and outlet plenums by a vertically oriented plate. The primary coolant, after leaving the reactor, enters the lower plenum and then flows upwards through the tube bundle area, the riser, in the cylindrical part of the SG. Water evaporation takes place on the secondary side of the riser. The tube bundle is the interface between the primary and secondary circuits. It is composed of U-shaped tubes with approximate height of 11 m. Water in the reactor coolant system thus flows first upwards and then downwards through the tube sheet, transferring the heat to the secondary fluid. The cooled down water is then poured into the other half of the header tank before flowing back to the reactor. Primary coolant enters the steam generator at 588–603 K on the hot-leg side and leaves at about 560 K on the cold-leg side. The steam generator is shown in **Figure 1** and the schematic drawing of the flow paths inside the RSG in **Figure 2**. The main RSG characteristics and parameters are shown in **Table 1**.

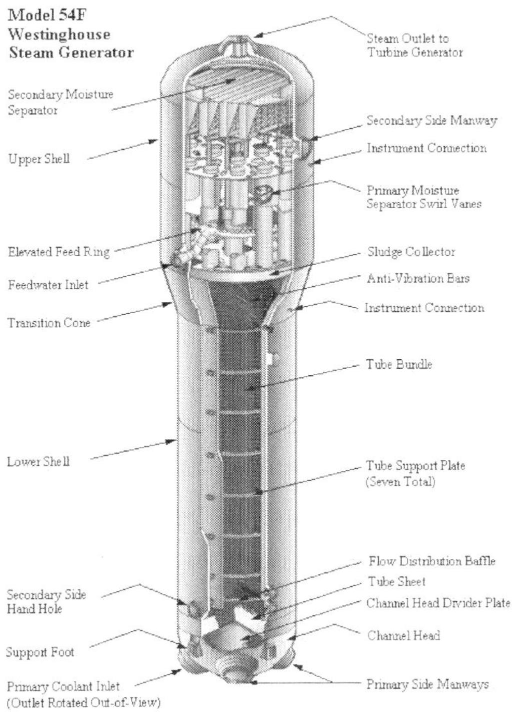

Figure 1. Recirculating steam generator (taken from Ref. [2]).

On the secondary side, feedwater enters the steam generator in the upper shell through a nozzle via a feed ring and is mixed with water draining from the moisture-separating equipment. Then, the water flows downwards through the downcomer to the tube sheet, a vertical plate separating the lower header and riser sections. The downcomer is situated between the tube bundle shroud and the SG outer shell. After reaching the tube sheet, water goes up, flowing below the shroud wall, in the central riser section where it is heated by the primary water flowing inside the U-tubes. Since the fluid in the secondary side, outside the tubes, is saturated, secondary-side water evaporates and a two-phase flow is established near the top of the tube bundle. The buoyancy force caused by a difference in densities of water inside the downcomer and the two-phase mixture in the tube bundle section ensures fluid circulation in the SG. As the secondary fluid flows upwards the riser, the steam quality increases up to 30%. This is not sufficient for a safe turbine operation as the droplets of water in the steam could damage the turbine blades. Thus, the steam that exits the tube bundle goes first into steam-moisture separators, composed of swirl vanes, and steam dryers, in the form of chevron separators, before it goes out of the steam generator through a nozzle at the top of the SG dome. The steam lines direct the steam flow into the turbine. The water collected by the separation devices falls down to the riser or is, for the most part, directed to the downcomer. After the drying process, the steam is saturated with a residual humidity, the moisture carryover, of less than 0.0025. Dry and saturated steam leaves the steam generator vessel and enters the steam lines.

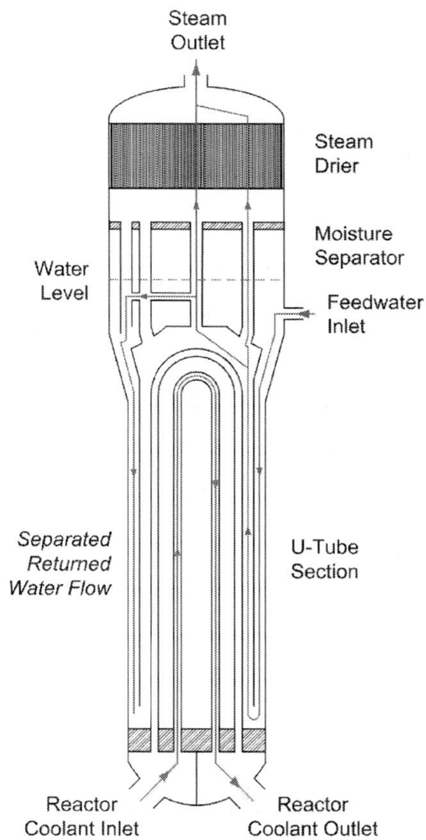

Figure 2. Simplified RSG flow paths.

Parameter	Value
Weight [tons]	330
Height [m]	20.7
Lower/upper diameter [m]	3.5/4.5
Heat transfer area [m^2]	7177
Number of U-tubes	5428
Tube inner/outer diameter [cm]	1.69/1.9
Power [MWt]	1000
Primary operating pressure [MPa]	15.5
Secondary operating pressure [MPa]	6.3
Primary inlet temperature [K]	598
Primary outlet temperature [K]	559
Secondary feedwater temperature [K]	492
Secondary steam temperature [K]	552
Primary coolant flow [kg/s]	4700
Secondary steam flow [kg/s]	545
Circulation ratio	3.7
Tubing material	Inconel 600

Table 1. Typical RSG dimensions and operating parameters.

About 25% of the secondary coolant is converted to steam on each pass through the generator, and the remainder is recirculated. That amount of recirculated coolant is described by the important design parameter called circulation ratio which is defined as the ratio between the total flow rate through the tube bundle and the flow rate of steam exiting the steam generator.

2.2. Regulation and control

In order to achieve the normal steam generator operation, a variety of parameters are subjected to regulation. Steam and feedwater flows need to be balanced; otherwise, the SG would overfill (if the feedwater flow is larger than the steam flow) or dry out (in the case the steam flow is larger than the feedwater flow). The amount of steam leaving the steam generator depends on the electrical load demand by the consumers. The turbine power is a linear function of the reactor coolant average temperature. The reactor temperature regulation system maintains the temperature by adjusting the position of the reactor control rods taking into account the signal of the turbine power. For the fast load changes, the excess steam is directed directly into the turbine condenser, thus bypassing the turbine. In that way, the steam generator pressure is being kept below the safety limits. Additionally, for the SG pressure control, power-operated relief and safety valves, mounted on the steam lines, are used. The relief valves are motor-operated valves, while safety valves are passive components.

The most important SG parameter subjected to regulation is the SG level. If the level is too low, the insufficient heat removal by the secondary side may cause evaporation of the reactor coolant, thus overheating of the reactor core. On the other hand, if the level is too high, the steam exiting the steam generator would carry water droplets (the void fraction would be higher than zero) which can be damaging to the turbine. The SG level is maintained by the feedwater flow by means of controller which continuously compares measured feedwater flow with steam flow and a compensated steam generator downcomer water level signal with a water level set point. A functional diagram of the steam generator level control system is shown in **Figure 3**.

The measured steam generator level is compensated by a lag controller ($1/(1+\tau_1 s)$) and subtracted from a desired reference SG level. That signal is then corrected by a proportional-integral (PI) controller ($K_1(1+1/\tau_2 s)$) and added to a difference between steam flow and feedwater flow signals. The resulting signal goes through a final PI correction ($K_2(1+1/\tau_3 s)$) before being used for feedwater flow control. Parameters K_1 and K_2 are scaling factors and τ_1, τ_2 and τ_3 time constants depending on the design of nuclear power plant control system.

According to **Figure 3**, in the case the reference level signal is larger than the measured level or the steam flow is larger than the feedwater flow, the feedwater flow will be increased by increasing the feedwater control valve area. In the opposite case, the control valve flow area will be decreased.

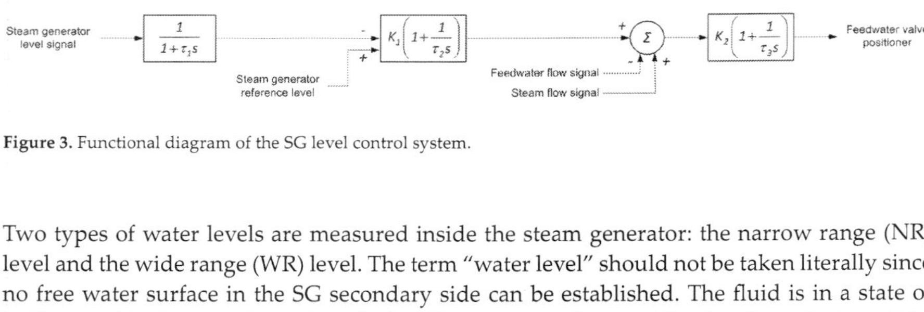

Figure 3. Functional diagram of the SG level control system.

Two types of water levels are measured inside the steam generator: the narrow range (NR) level and the wide range (WR) level. The term "water level" should not be taken literally since no free water surface in the SG secondary side can be established. The fluid is in a state of boiling, and in the area above the tube bundle, steam quality steadily rises from the top of the U-tubes to the inlet in the steam separators. Thus, the level is deduced from the pressure difference, pressures being measured at two different heights. The level is affected by variations in the fluid density as well as residual pressure drops.

In general, the SG level is a measure of a pressure difference inside the steam generator compared to a pressure difference between the liquid and gas phases. It is calculated by the expression:

$$SGLVL[\%] = 100 \cdot \frac{\Delta p - \Delta p_{0\%}}{\Delta p_{100\%} - \Delta p_{0\%}} \tag{1}$$

where $\Delta p_{0\%} = \rho_g gh$ and $\Delta p_{100\%} = \rho_l gh$.

The height in the expressions for the pressure difference is the distance between the measurement taps. For both the narrow and wide range measurements, the upper tap is in the separator area. The lower tap for the NR level measurement is just below the U-tube bend area, near the top of the downcomer. For the WR level measurement, the lower tap is at the bottom of the downcomer. The total height for the NR level measurement is about 5 m and for the wide range 15 m.

The narrow range level is used to control the feedwater flow rate. Except in the case of extreme events, such as very fast transients caused by accidental depressurization, the narrow range level is maintained constant by the control system. The feedwater flow rate and temperature, and thermal hydraulic SG conditions, have much larger influence on the wide range level. Therefore, the WR level is only used as a level indicator for the NPP operators during slow running transients or during the plant shutdown and start-up operation modes after an outage. Dependence of the WR level on the dynamic SG pressure prevents it to be used for the control of the NPP performance [4].

3. Numerical analysis and performance results of the RSG

The mathematical simulation of the SG normal and transient operation and behaviour was performed with system computer code RELAP5 [5] on the example of a two-loop PWR power plant Krško [6].

Correlations for the heat transfer calculation in steam generators are given first. Then, SG modelling requirements and mathematical models of the plant are presented, followed by analytical results of SG performance in accident conditions during the loss of the NPP heat sink.

3.1. Physical models of thermal hydraulic and heat transfer phenomena

3.1.1. Thermal hydraulic conditions

The RELAP5 code is a six-equation one-dimensional, nonhomogeneous, nonequilibrium transient system code. It solves mass, momentum and energy conservation differential equations for the two phases, gas and liquid, hence, the six conservation equations. The equations will not be presented here because they are standard fluid conservation equations, although they include many fine transport mechanisms in order to realistically simulate thermal hydraulic system behaviour [5]. For example, momentum equations take into account wall friction, momentum transfer due to interface mass transfer, interface frictional drag and force due to virtual mass. The interface mass and heat transfer terms are also treated by the mass and energy conservation equations.

Efficient and reliable SG operation requires efficient steam separation. Separators must be capable of achieving very low moisture carryover. High carryover will result in turbine efficiency losses as well as the potential for turbine blade erosion. Efficient steam separator design also requires that the primary separation stage has the low pressure drop and low steam carryunder in the downcomer flow in order to support efficient recirculation through the tube bundle. Furthermore, to allow flexibility in water level operation, the separators must be able to operate over a wide range of water levels.

The computational separator model consists of a special hydraulic volume component with junction flows (**Figure 4**). A steam-water inflowing mixture is separated by defining the quality of the outflow streams using empirical functions. The void fraction of the vapour at the separator outlet junction J_1 depends on the void fraction in the separator thermal hydraulic volume according to the left curve in **Figure 5**. If the vapour void fraction in the separator volume is larger than the input parameter, labelled as VOVER, the outlet void fraction will be 1.0, and pure gas vapour will be released out of the separator. If the separator vapour void fraction is less than the value of VOVER, then the outflow is going to be a two-phase mixture of gas and liquid. Thus, changing the VOVER parameter, the code user can control the state of fluid leaving the separator. In the case of VOVER parameter being small, the separator is going to act as an ideal separating device, discharging pure vapour, and in the case of VOVER having a high value, close to 1.0, the separator component will perform as a standard junction releasing fluid in the same state as is entering the separator.

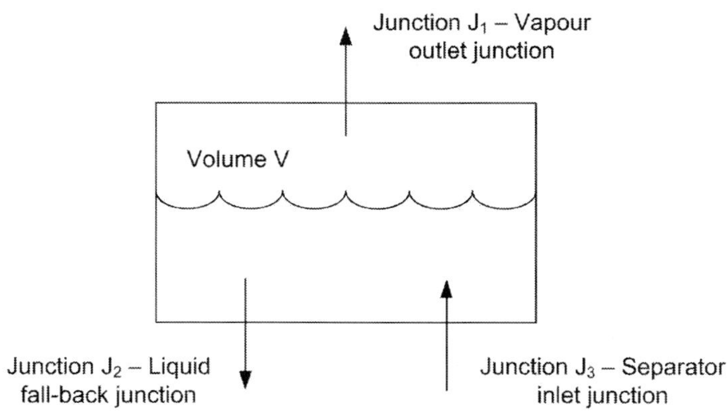

Figure 4. Typical separator volume and junctions.

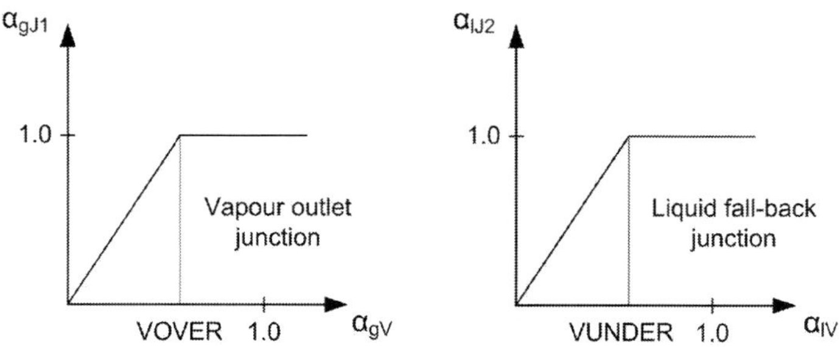

Figure 5. Outlet junctions void fractions.

The flow of the separator liquid drain junction is modelled in a manner similar to the steam outlet except that pure liquid outflow is assumed when the volume liquid void fraction is greater than the value of VUNDER (**Figure 5**). Typical values of VOVER and VUNDER parameters are 0.001 and 0.1, respectively.

3.1.2. Correlations for the heat transfer calculation

Steam generator operation depends on the wall-to-fluid heat transfer on the secondary side. During the steady-state operation, the steam generator water level is constant, but during the transient, it can vary in the large range; thus, both convective and boiling heat transfers occur across the U-tubes. The total wall heat flux is composed of convective heat transfer to gas and liquid phases, boiling and condensation heat fluxes. During boiling, the saturation temperature based on the total pressure is the reference temperature, and during condensation, the saturation temperature based on the partial pressure is the reference value. The expression for a heat flux is given as

$$q_{\text{wall\_total}} = h_{g,g}\left(T_w - T_g\right) + h_{g,spt}\left(T_w - T_{spt}\right) + h_{g,spp}\left(T_w - T_{spp}\right) + h_{l,l}\left(T_w - T\right)_l + h_{l,spt}\left(T_w - T_{spt}\right), \tag{2}$$

where

- $h_{g,g}$ is the heat transfer coefficient (HTC) from the wall to the gas phase, and the reference temperature is the gas temperature.
- $h_{g,spt}$ is the HTC from the wall to the gas phase, and the reference temperature is the saturation temperature at the total fluid pressure.
- $h_{g,spp}$ is the HTC from the wall to the gas phase, and the reference temperature is the saturation temperature at the partial pressure of steam.
- $h_{l,l}$ is the HTC from the wall to the liquid phase, and the reference temperature is the liquid temperature.
- $h_{l,spt}$ is the HTC from the wall to the liquid phase, and the reference temperature is the saturation temperature at the total fluid pressure.
- T_w, T_g and T_l are the wall surface, gas and liquid temperatures, respectively.
- T_{spt} and T_{spp} are the saturation temperatures at the total fluid pressure and at the partial pressure of steam, respectively.

Correlations used to calculate heat transfer in the steam generators, depending on the heat transfer regimes, are given in **Table 2**.

For the convective heat transfer, the correlation is given by Churchill and Chu [7]:

$$Nu = \left(0.825 + \frac{0.387 Ra^{1/6}}{\left(1 + \left(\frac{0.492}{\mathrm{Pr}}\right)^{9/16}\right)^{8/27}}\right)^2, \tag{3}$$

where Nu, Ra and Pr are Nusselt, Rayleigh and Prandtl numbers, respectively.

Heat transfer phenomena	Correlation
Convection to noncondensable steam-water mixture, supercritical, single-phase liquid or gas flows	Churchill and Chu [7]
Subcooled or saturated nucleate boiling	Forster and Zuber [8]
Subcooled or saturated transition boiling	Chen et al. [9]
Subcooled or saturated film boiling	Bromley [10]
Condensation heat flow	Nusselt [11]

Table 2. Wall heat transfer correlations.

The Forster-Zuber correlation [8] for the nucleate boiling heat transfer coefficient is given as

$$h = 0.00122 \frac{k_l^{0.79}\, c_{pl}^{0.45}\, \rho_l^{0.49}\, g^{0.25}}{\sigma^{0.5}\, \mu_l^{0.29}\, h_{lg}^{0.24}\, \rho_g^{0.24}} \Delta\, T_w^{0.24}\, \Delta\, p^{0.75}, \tag{4}$$

where k is the thermal conductivity, c_p the specific heat, ϱ the density, g the gravity acceleration, σ the surface tension, μ the viscosity, h_{lg} the enthalpy of boiling, ΔT_w the difference between the wall and fluid temperatures and Δp the difference between the saturation and total pressures.

The Chen transition boiling model [9] considers the total transition boiling heat transfer to be the sum of individual components, one describing wall heat transfer to the liquid and a second describing the wall heat transfer to the vapour. The model is expressed as

$$q = q_l A_l + 0.0185\, \mathrm{Re}^{0.83}\, \mathrm{Pr}^{1/3} (T_w - T_g)(1 - A_l), \tag{5}$$

where q is the transition boiling heat flux and A_l is the fractional wall-wetted area.

The Bromley [10] correlation for the heat transfer coefficient during film boiling is given as

$$h = 0.62\alpha \left(\frac{g \rho_g k_g^2 (\rho_l - \rho_g)\, h_{lg}\, c_{pg}}{L (T_w - T_{spt})\, \mathrm{Pr}_g} \right)^{0.25}, \tag{6}$$

where α is the void fraction and L the characteristic length.

Finally, the Nusselt [11] correlation for the condensation heat transfer coefficient is given as

$$h = 1.1006\, k_l \left(\frac{g \rho_l \Delta \rho}{\mu_l^2\, \mathrm{Re}_l} \right)^{1/3}. \tag{7}$$

3.2. Numerical simulation: the NPP model and results

3.2.1. The RELAP5 computational model of the power plant

The nodalization scheme (mathematical model) of the two-loop PWR nuclear power plant, analyzed herein, is shown in **Figure 6**. The boxes represent hydraulic control volumes (CVs) connected by junctions. The reactor is in the middle of the figure, connected by pipes with two steam generators (SG1 and SG2). The scheme also includes other important NPP components and systems: the pressurizer; the safety injection system; the main feedwater; and auxiliary feedwater systems, steam lines, etc.

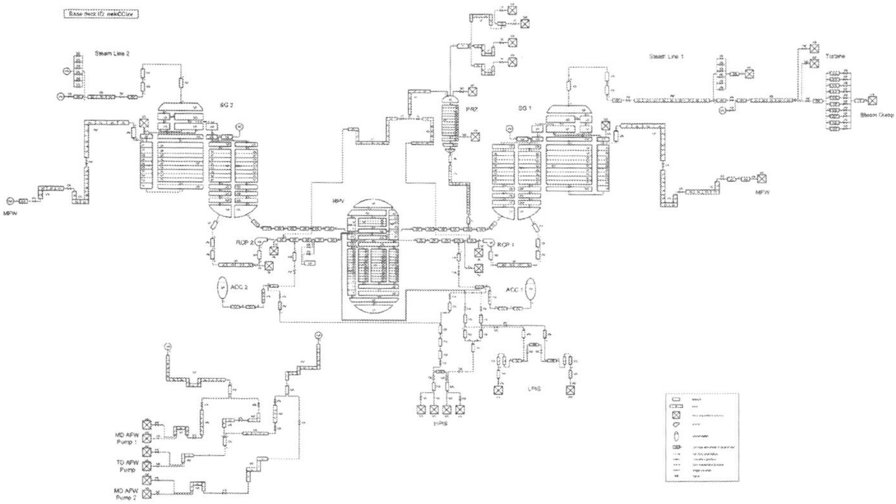

Figure 6. RELAP5 nodalization of the NPP Krško.

Regarding the steam generator model, on the primary side, the inlet plenum is represented with two control volumes (215 and 217), the tube sheet inlet with one CV (219), the U-tube section with 20 control volumes (upward part of U-tubes—CVs 223, 22501–22508, 227, and downward part of U-tubes—CVs 233, 23501–23508, 237), the tube sheet outlet with one CV (241) and the outlet plenum with two CVs (243 and 245).

On the secondary side, the downcomer is represented with 11 control volumes (411, 41301–41310), the riser section also with 11 CVs (415, 41701–41709+-, 419), the separator with one CV (421), the steam plenum with dryer with one CV (423), the separator bypass region with two CVs (425, 427) and, finally, the steam generator dome with one control volume – 429. This is the model of the SG1. The SG2 has the same model with the different numbering.

3.2.2. Steady-state calculation

The first step in the NPP and the SG model development is the qualification of the code nodalization. This is done by comparing plant's main operating parameters with computer steady-state simulation at full power. Parameters of interest are primary and secondary system pressures, reactor coolant, feedwater and steam mass flow rates, reactor coolant temperatures, pressurizer and steam generator liquid levels, primary and secondary system fluid masses, heat transfer inside SGs, circulation ratio, etc. If the calculated values differ less than approximately 1% than the real plant data, we can say that nodalization is qualified for the plant safety analyses.

The steam generator qualification process of the RELAP5 model includes calculation of pressures, temperatures, fluid flows and liquid levels inside the SG. Additionally, geometrical representation of the computational model and calculation of SG conditions at partial loads need also to correspond to manufacturer data, as shown in **Figures 7** and **8**, respectively. The SG pressure at a lower load is higher than the pressure at a higher load. As the pressure increases, water evaporates at a slower rate, and a total SG secondary fluid mass is 50% higher at a 10% load than at a 100% load. The steam flow at a full load is 540 kg/s, while at a 10% load, the flow is only 40 kg/s, achieved by increase of pressure and decrease in feedwater temperature. The pressure difference of 1 MPa, as observed in **Figure 8**, results in steam temperature change of only 10 K. The highest impact of load change is on the circulation ratio which decreases from the value of 41 at a 10% load to a value of 3.7 at a full load.

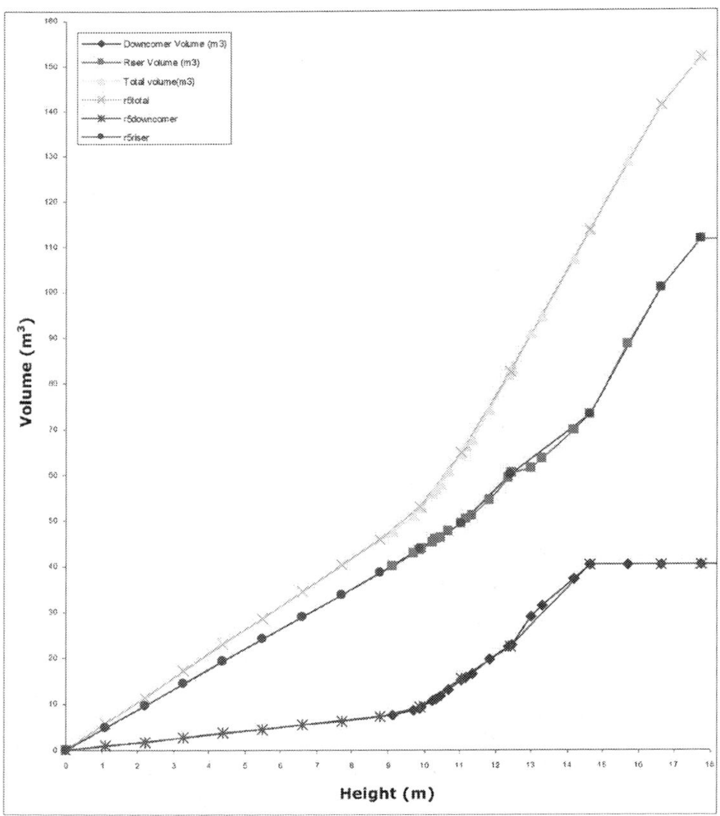

Figure 7. SG secondary side volume versus height.

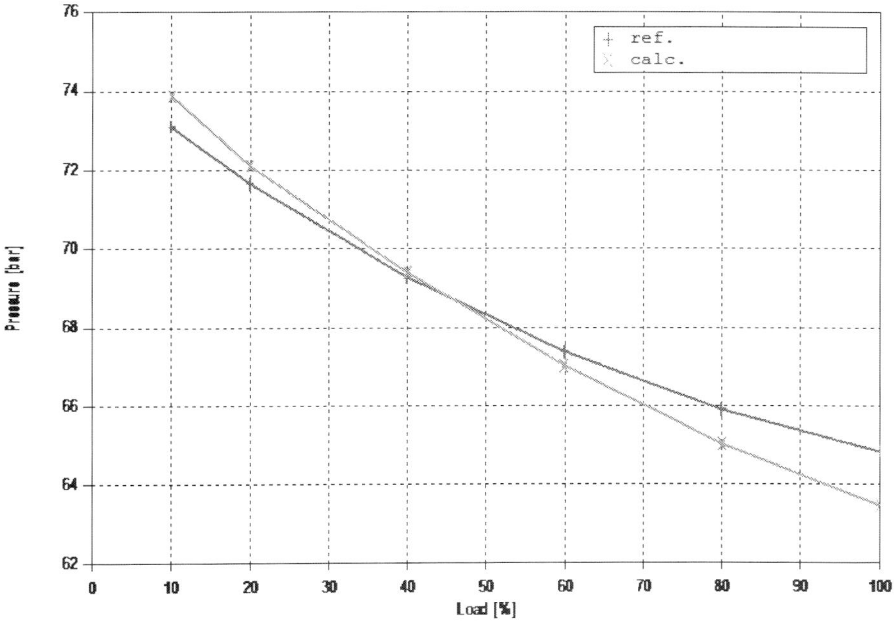

Figure 8. SG steam pressure versus load.

3.2.3. Transient calculation

In order to illustrate SG dynamic behaviour, a representative accident of the loss of all electrical power was selected. The unavailability of AC power supply to NPP systems means that important operational and safety components, such as big pumps which provide driving force to primary and secondary system coolant flows, will not work. There will not be feedwater flow into the SG, and the water level will decrease. On the primary side, the reactor core will heat up and subsequently, if in the meantime, no power restoration occurs, melt. Plant conditions will be further complicated by the loss of reactor coolant through damaged reactor coolant pump seals. The seals are normally cooled by the high-pressure water flow provided by the charging pump which, without electrical motor drive, does not operate.

There are two ways of mitigating accident consequences. First, there is a passive steam-driven auxiliary feedwater (AFW) pump that can provide water for the cooling of steam generators. The steam is taken directly from the SG outlet. In addition, the plant operator can reduce pressure by controlling SG relief valves and prolong the time to the core damage because of increased cooling of the reactor coolant system (RCS). Power supply needed for those actions is provided by the accumulator batteries installed at the plant. The three scenarios ((1) no AFW flow, no SG depressurization; (2) AFW flow, no SG depressurization; (3) AFW flow, SG depressurization) were analyzed according to the aforementioned mitigating options.

The AFW system water injection provided secondary-side heat sink. The SG pressure was maintained at 8 MPa by the operation of SG safety valves (**Figure 9**). Natural circulation was established in the primary system after the stoppage of coolant pumps, heat source being the core decay heat and heat sink of the steam generators. Primary system water evaporated in the core, and steam condensed in the SG U-tubes. The heat (**Figure 10**) was transferred to the secondary-side boiling water which level (**Figure 11**) was maintained by injection from the AFW system. Oscillatory behaviour was due to operation of safety valves, and continuous short-term steam releases to keep pressure at 8 MPa. Large condensate storage tank (CST) water inventory (860 m^3) provided AFW flow for almost 70 h. Depletion of CST inventory led to dry out of the SGs. The CST tank could be filled up at any time during the accident, but in this conservative analysis, no provision of maintaining the CST water inventory was taken into account. Soon after the CST depletion, the RCS was heated up, water boiling in the core accelerated and the core started to uncover. If there is no immediate action to inject water in the core, the core will melt.

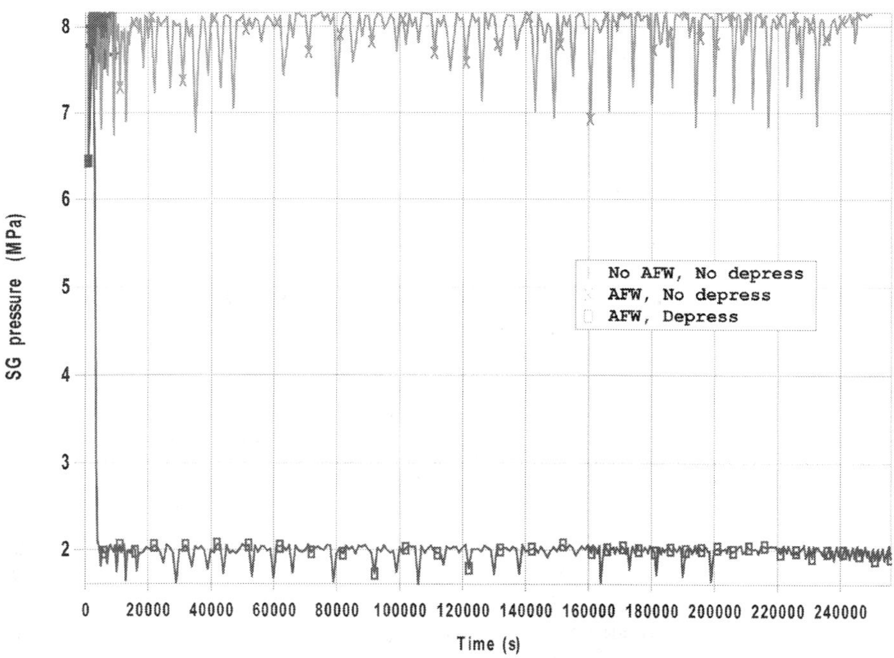

Figure 9. SG pressure.

http://dx.doi.org/10.5772/66962

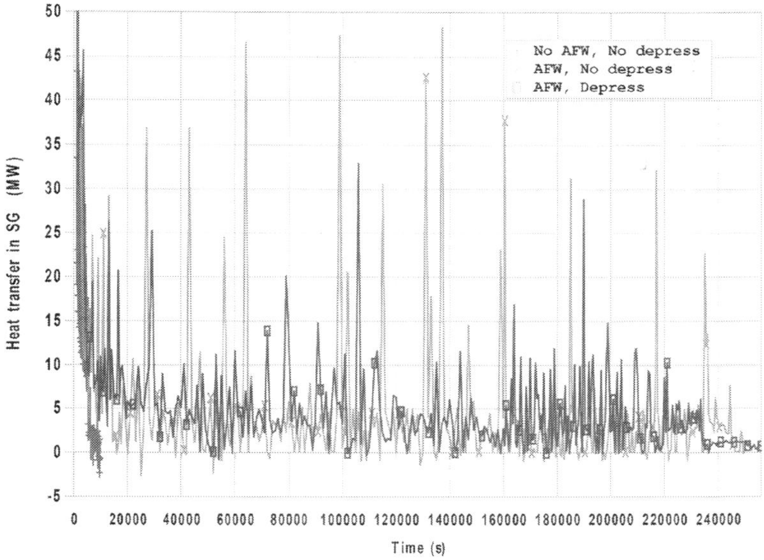

Figure 10. Heat transfer in steam generators.

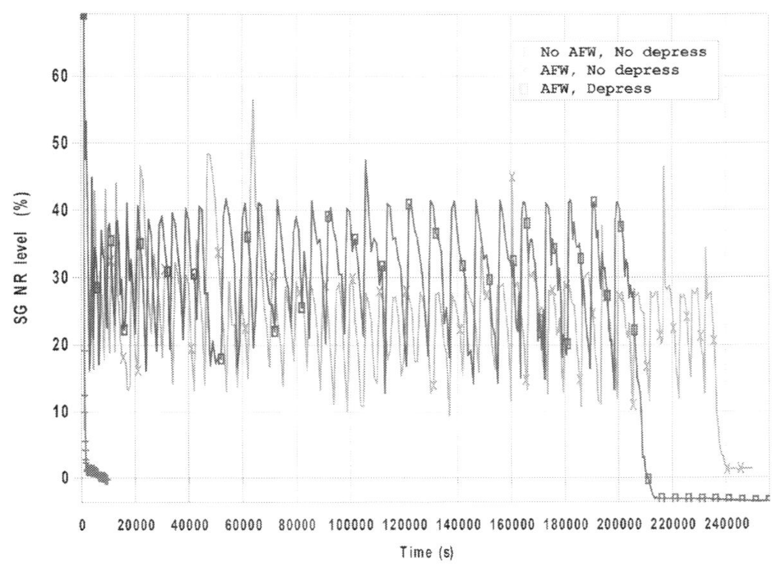

Figure 11. SG narrow range water level.

Operator action to rapidly depressurize secondary side to 2 MPa using SG relief valves led to fast primary-side cooldown and depressurization. The primary and secondary systems were closely coupled with little difference in pressures and temperatures. Both systems were in saturated conditions, temperature depending on the saturation pressure. Since the core decay heat was only a couple of percents of the total power and the heat transfer area in the SG was very large, the temperature difference was only few kelvins, and, thus, pressures were almost the same. With the operator action, the CST was emptied at ~202,500 s. That was 30,000 s earlier than in the case without any operator actions due to higher AFW flow required during SG steam extraction to satisfy prescribed RCS cooldown rate. Reducing the primary pressure to 2 MPa enabled the actuation of accumulators' water injection which quenched the core. Therefore, although secondary-side heat sink was lost earlier, the larger RCS water inventory increased the time to core damage. In the case with no AFW flow, the core heat-up started after 2.2 h. In the case with the AFW availability, the core temperatures started to increase after 65 h for the case without SG depressurization, and after 70 h when the operator controlled the SG pressure.

4. Helical-coil steam generators

4.1. Description and main characteristics

Many future nuclear reactor projects, especially innovative small- and medium-sized reactor systems [12], are expected to use helically coiled pipes for the steam generators. Their favourable characteristics justify the helical tube SG development in the nuclear field. In particular, helical tubes provide enhanced heat and mass transfer rates, a higher critical heat flux during boiling and evaporation and a better capability to accommodate the thermal expansion, in addition to allow a more compact design of the SG. Helical coils are particularly attractive for small and medium modular reactors (SMRs) since many of them adopt an integral layout. Compactness and efficiency improvement become particularly important for this type of reactors, as all the primary system components are located inside the reactor vessel.

The helical coil SG design and operation will be explained on the example of the IRIS reactor [13]. The international reactor innovative and secure (IRIS), an integral, modular, medium-sized (335 MWe) PWR, has been under development since the turn of the century by an international consortium led by Westinghouse and including over 20 organizations from nine countries. IRIS features an integral vessel that contains all the major reactor coolant system components, including the reactor core, pumps, the steam generators and the pressurizer. The unique IRIS safety-by-design approach provides a very powerful first level of defence in depth approach by eliminating accidents, at the design stage, or decreasing their consequences and probabilities when outright elimination is not possible. There are no primary system pipings, and a large-break loss-of-coolant accident (LOCA), related to a double-ended break of a primary leg pipe, is avoided. The passive safety systems increase plant safety by providing core decay heat removal during accident conditions even where there is no electrical power supply.

Steam generators are located in the space between the core barrel, precisely the shroud enclosing the riser section, and the reactor vessel wall. There are eight SG units in total, designed as once-through heat exchanging units (**Figure 12**). They are made of helical tubes with

http://dx.doi.org/10.5772/66962

secondary fluid flowing inside the tubes. The feedwater header is located at the bottom of the SG module, while the steam header is positioned at the top of the steam generator. The tubes are wrapped around the inner supporting column. The primary cooling water flows outside of the tubes, through the tube bundle. The primary reactor coolant pumps are installed above the steam generators and drive coolant from the top to the bottom of the SG. Thus, a counter-current flow regime is developed inside the SG, the primary coolant flows from the top to the bottom of the SG and the secondary fluid flows in the opposite direction.

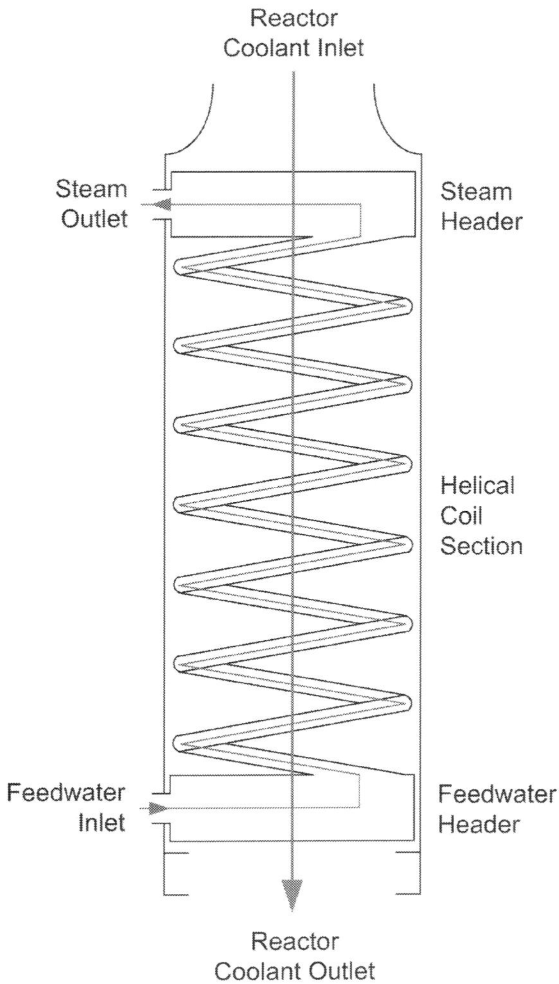

Figure 12. Simplified IRIS helical-coil SG flow paths.

The tubes are set up in annular rows and connected to steam and feedwater headers (**Figure 13**), which, on the other hand, are connected to feedwater and steam lines piping by nozzles mounted on the external surface of the reactor vessel wall. At the tube inlet, orifices which reduce fluid flow are installed in order to maintain uniform flow distribution across the tubes and to prevent parallel channel flow instabilities. The pressure drops at these orifices are of the same order of magnitude as the pressure drops in the tubes. The main IRIS reactor helical-coil SG characteristics and parameters are shown in **Table 3**.

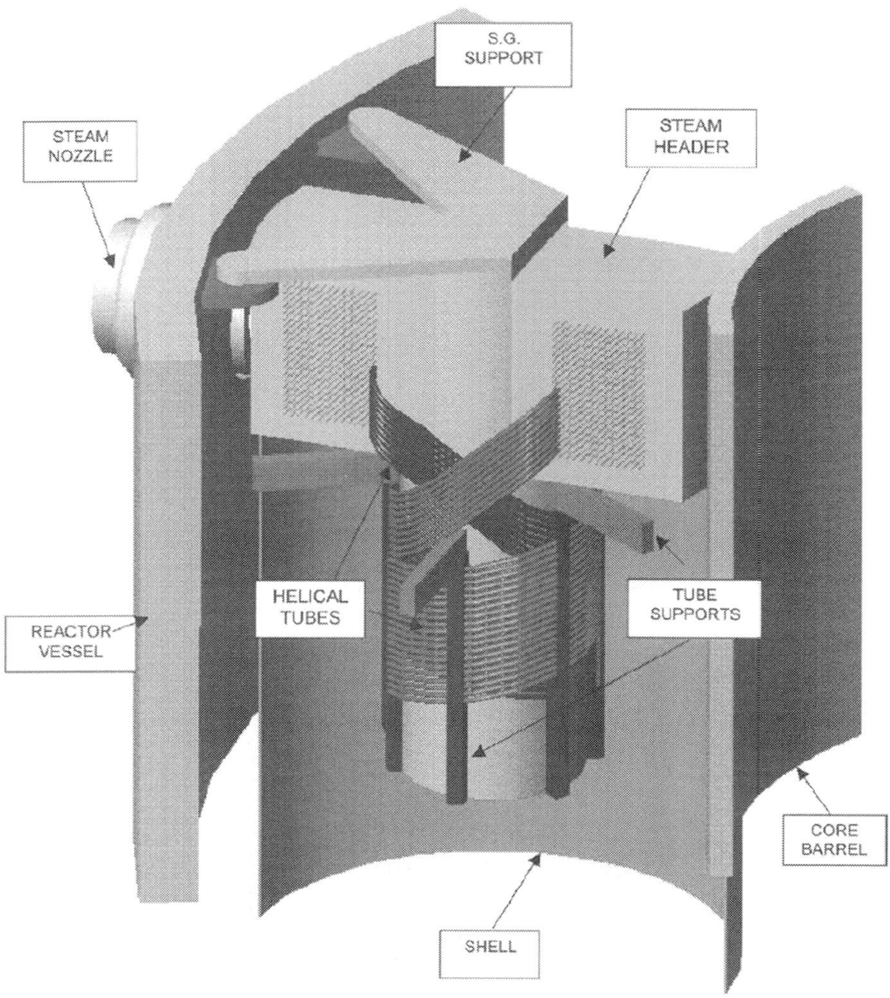

Figure 13. IRIS SG steam header (taken from Ref. [13]).

Parameter	Value
Height [m]	8.5
External shell inside diameter [m]	1.62
Internal shell outside diameter [m]	0.61
Number of helical rows	21
Number of tubes	655
Tube bundle average length [m]	32
Tube inner/outer diameter [cm]	1.32/1.75
Power [MWt]	125
Primary operating pressure [MPa]	15.5
Steam outlet pressure [MPa]	5.8
Primary inlet temperature [K]	602
Primary outlet temperature [K]	565
Secondary feedwater temperature [K]	497
Secondary steam temperature [K]	590
Primary coolant flow [kg/s]	589
Secondary steam flow [kg/s]	62.5
Tubing material	Ni-Cr-Fe alloy

Table 3. IRIS helical-coil SG dimensions and operating parameters.

Primary and secondary fluid temperature profiles are shown in **Figure 14**. The primary fluid is water, and its temperature continuously increases along the height of the SG (decreases in the flow direction). On the secondary side, subcooled water enters the tubes, heats up, boils and produces superheated steam in the upper part of the tubes. During boiling, steam temperature slightly decreases due to a small pressure drop across the tubes. The total pressure drop on the secondary side, from the feedwater inlet to the steam outlet, is 323 kPa and in the boiling part of the tubes 140 kPa. The void fraction at point when superheating starts is 0.98.

4.2. Computational model and numerical simulation of the IRIS reactor

The RELAP5 code was used to model the facility and simulate accident conditions. Although the code lacks appropriate correlations for the heat transfer coefficient in helical pipes, comparison with detailed calculation [14] shows a good agreement between the results. The worldwide experience and quality models inside the RELAP5 were the reasons for selecting it to perform IRIS preliminary safety assessment studies [15].

The nodalization of the IRIS nuclear power plant is shown in **Figure 15**. On the left side of the figure, the reactor core is in the lower part (CV 110), the riser in the middle part (CVs

120–123) and pressurizer in the upper part (CV 130). On the right side, the pump (CV 191) is below the pressurizer and connected to the steam generator (CV 211). The SG annular dead space surrounding the SG shell and the inner inactive region are modelled as CV 240 and CV 241, respectively. The lower downcomer region (CV 101) is below the steam generator. The secondary side of the steam generator is modelled with CV 271. All eight steam generators are modelled, but only one is presented in the figure. The primary side of the SG was represented with 25 control volumes and the secondary side with 50 control volumes. That fine nodalization was obtained during the optimization process that resulted with stable and correct steady-state performance. Parallel channel flow instabilities were eliminated by modifying pressure loss coefficients. The heat transfer coefficient was also adjusted using multiplication factor to accommodate the inability of the RELAP5 to model curved helical-coil geometry. In that geometry eddy flow currents are created inside the coils, which promote mixing of the fluid and, thus, increase heat transfer capability at the expense of higher pressure drops [16].

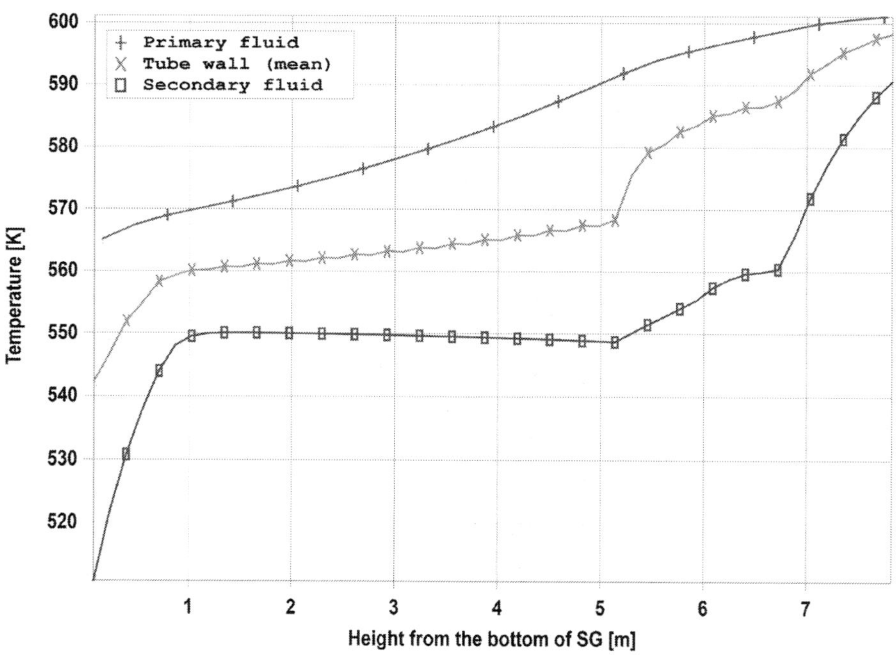

Figure 14. Primary and secondary fluid temperature profiles in the IRIS NPP.

http://dx.doi.org/10.5772/66962

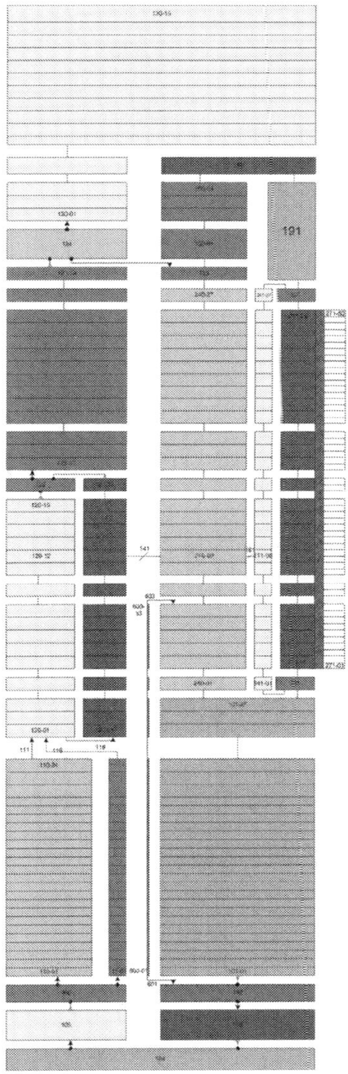

Figure 15. RELAP5 nodalization of the IRIS reactor vessel.

A small-break LOCA was analyzed to demonstrate steam generator behaviour in accident conditions. The initiating event was the rupture of a chemical and volume control system pipe connected to the upper annular pump suction plenum of the reactor vessel. IRIS is designed to limit the loss of coolant from the vessel rather than relying on systems to inject water into the reactor vessel. It has a compact, small diameter, high-design pressure containment that assists in limiting the blowdown from the reactor vessel by providing a higher back pressure in the initial stage of the accident. Furthermore, four trains of passive emergency heat removal systems (EHRS) help to depressurize the primary system by condensing steam, coming out of the reactor core, on the steam generators tubes (depressurization without the loss of mass), and to remove the decay heat. Finally, it features automatic depressurization system to condense steam, released from the top of the pressurizer, in the pressure suppression pool located inside the containment.

Following the initiating event, the LOCA mitigation signal is rapidly actuated, the reactor and reactor coolant pumps are tripped and the four EHRS subsystems are actuated by closing the main feed and steam isolation valves and by opening the fail-open valves in the EHRS return lines connected to the SG feed lines. The EHRS is composed of pipes, valves and heat exchangers submerged in the water tank outside the containment. After the initial decrease of the coolant inventory in steam generators caused by the isolation of the feedwater flow, the EHRS operation restores the SG secondary fluid mass (**Figure 16**) and enables heat removal out of the steam generators (**Figure 17**). It does not take long for the situation to stabilize to ensure safe reactor conditions. Equalization of reactor vessel and containment pressures marks the end of the blowdown phase and start of a long-term cooling phase by the continued operation of the EHRS, with the pressure being slowly reduced as the core decay heat decreases.

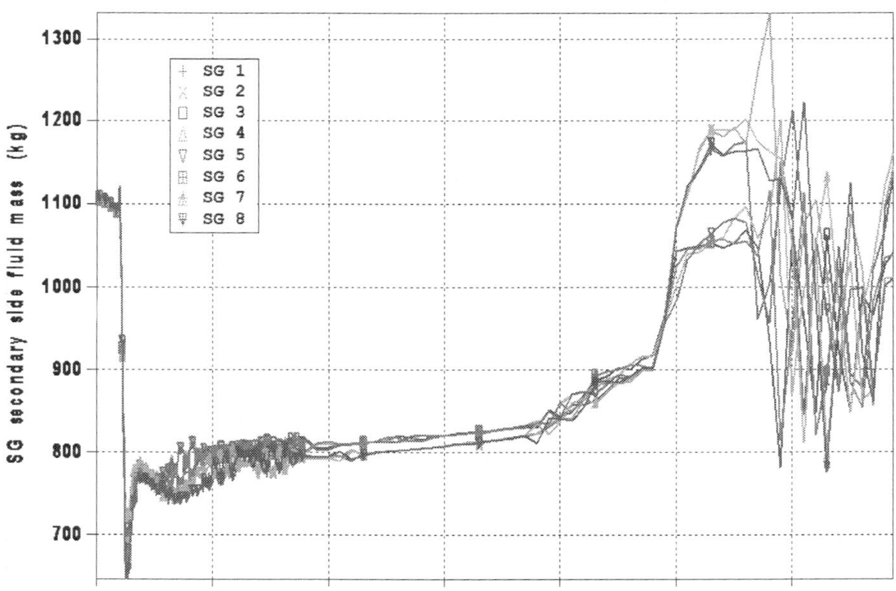

Figure 16. SG secondary-side fluid mass in IRIS NPP.

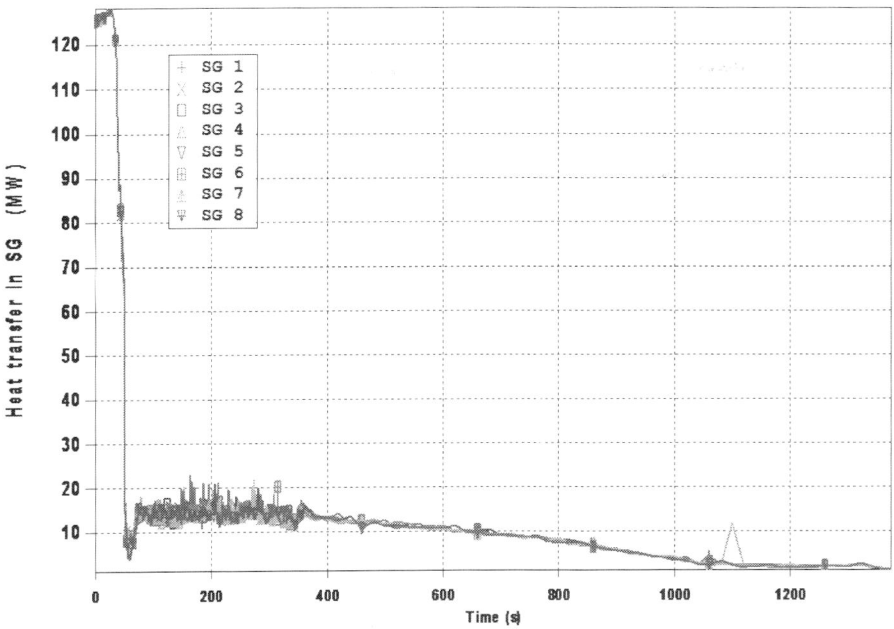

Figure 17. Heat transfer in IRIS NPP steam generators.

5. Conclusions

Steam generators (SGs) are nuclear power plant components where the steam, which drives the turbine, is produced. They also represent barrier that prevents radioactive fission products to escape outside the containment building.

In order to ensure safe operation of a nuclear power plant, SG parameters, the steam flow, steam pressure and temperature, feedwater temperature, circulation ratio and total inventory mass, have to be maintained within prescribed values. These values depend on the operating window and the thermal load. The pressure is controlled by the relief and safety valves and the inventory by the feedwater flow.

The inverted U-tube steam generators have quite large water mass in the secondary side which is important during accidents of loss of the secondary heat sink. In that design, primary system water which is at a higher pressure than the secondary fluid flows inside the tubes, while the secondary fluid is on the outer side. In the helical-coil SGs, the situation is opposite: the primary water is on the tubes' outside surface keeping the tubes in the state of compression. The loss of the secondary flow is here more critical due to limited amount of water, but this is compensated with passive auxiliary safety systems.

In the event of losing the primary coolant pumps, the natural circulation between the reactor core and the SGs can ensure safe cooling of the reactor core by establishing the two-phase flow inside the tubes, with only minimal operator actions required. Thus, the design of the nuclear steam generators and the auxiliary systems enables safe NPP operation, not only during the normal plant operation but also during the accident conditions.

Author details

Siniša Šadek* and Davor Grgić

*Address all correspondence to: sinisa.sadek@fer.hr

University of Zagreb Faculty of Electrical Engineering and Computing, Zagreb, Croatia

References

[1] Bonavigo L, De Salve M. Issues for Nuclear Power Plants Steam Generators. In: Dr. Uchanin V, editor. Steam Generator Systems: Operational Reliability and Efficiency. Rijeka InTech; 2011. pp. 371–392. DOI: 10.5772/14853

[2] IAEA, editor. Assessment and Management of Ageing of Major Nuclear Power Plant Components Important to Safety: Steam Generators. Technical Reports Series no. IAEA-TECDOC-1668: ed. Vienna: International Atomic Energy Agency; 2011. 273 p.

[3] Cinotti L, Bruzzone M, Meda N, Corsini G, Lombardi CV, Ricotti M, Conway LE. Steam Generator of the International Reactor Innovative and Secure. In: Proceedings of ICONE-10, 10th International Conference on Nuclear Engineering; April 14–18, 2002; Arlington, VA. 2002. pp. 1–8.

[4] Girard S Physical and Statistical Models for Steam Generator Clogging Diagnosis. 1st ed. eBook: Springer International Publishing; 2014. 97 p. DOI: 10.1007/978-3-319-09321-5

[5] U.S. NRC, editor. RELAP5/MOD3.3 Code Manual Volume I: Code Structure, System Models, and Solution Methods. NUREG/CR-5535/Rev 1 ed. Washington, DC: U. S. Nuclear Regulatory Commission; 2001. 428 p.

[6] Grgić D, Špalj S, Bajs T. Main Steam Line Break—Hot Full Power Analysis for NPP Krško Using RELAP5/PARCS Coupled Code. In: Čavlina N, Pevec D, Bajs T, editors. Conference Proceedings of the 6th International Conference on Nuclear Option in Countries with Small and Medium Electricity Grids; May 21–25, 2006; Dubrovnik, Croatia. Zagreb, Croatia: Croatian Nuclear Society; 2006. pp. S6.74.1–S6.74.14.

[7] Churchill SW, Chu HHS. Correlating Equations for Laminar and Turbulent Free Convection from a Vertical Plate. International Journal of Heat and Mass Transfer. 1975;**18**(11):1323–1329.

[8] Forster HK, Zuber N. Dynamics of Vapor Bubbles and Boiling Heat Transfer. AIChE Journal. 1955;**1**(4):531–535.

[9] Chen JC, Sundaram RK, Ozkaynak FT, editors. A Phenomenological Correlation for Post-CHF Heat Transfer. NUREG-0237 ed. Washington, DC: U.S. Nuclear Regulatory Commission; 1977. 144 p.

[10] Bromley LA. Heat Transfer in Stable Film Boiling. Chemical Engineering Progress. 1950;**46**:221–227.

[11] Nusselt W Die oberflächenkondensation des wasserdampfes. Zeitschrift des Vereines Deutscher Ingenieure. 1916;**60**(27):541–546.

[12] IAEA, editor. Innovative Small and Medium Sized Reactors: Design Features, Safety Approaches and R&D Trends. Technical Reports Series no. IAEA-TECDOC-1451 ed. Vienna: International Atomic Energy Agency; 2005. 214 p.

[13] Carelli MD, Conway LE, Oriani L, Petrović B, Lombardi CV, Ricotti ME, Barroso ACO, Collado JM, Cinotti L, Todreas NE, Grgić D, Moraes MM, Boroughs RD, Ninokata H, Ingersoll DT, Oriolo F. The Design and Safety Features of the IRIS Reactor. Nuclear Engineering and Design. 2004;**230**:151–167. DOI: 10.1016/j.nucengdes.2003.11.022

[14] Caramello M, Bertani C, De Salve M, Panella B. Helical Coil Thermal Hydraulic Model. Journal of Physics: Conference Series. 2014;**547**:1–10. DOI: 10.1088/1742-6596/547/1/012034

[15] WEC LLC, editor. IRIS Preliminary Safety Assessment. WCAP-16082-NP Westinghouse Non-Proprietary Class 3 ed. Pittsburgh, PA: Westinghouse Electric Company LLC; 2003. 142 p.

[16] Cioncolini A, Cammi A, Cinotti L, Castelli G, Lombardi C, Luzzi L, Ricotti ME. Thermal Hydraulic Analysis of IRIS Reactor Coiled Tube Steam Generator. In: American Nuclear Society Topical Meeting in Mathematics & Computations; April 6–11, 2003; Gatlinburg, TN. LaGrange Park, IL: American Nuclear Society; 2003. pp. 1–9.

Chapter 9

Industrial Heat Exchanger: Operation and Maintenance to Minimize Fouling and Corrosion

Teng Kah Hou , Salimnewaz Kazi ,
Abu Bakar Mahat , Chew Bee Teng ,
Ahmed Al-Shamma'a and Andy Shaw

Additional information is available at the end of the chapter

http://dx.doi.org/10.5772/66274

Abstract

Heat exchanger is equipment used to transfer heat from one fluid to another. It has extensive domestic and industrial applications. Extensive technical literature is available on heat exchanger design, operation and maintenance, but it is widely scattered throughout the industrial bulletins, industrial design codes and standard, technical journals, etc. The purpose of this book chapter is to consolidate into basic background and concepts design of heat exchangers, operation, cleaning and green technology maintenance on heat exchanger closely related to the industrial practices.

Keywords: heat exchanger, fouling, fouling mitigation, green technology, cleaning of heat exchangers

1. Introduction

Heat exchanger plays an important role in industrial application. It is implemented for the purposes of heating and cooling of large-scale industrial process fluids [1]. Heat exchanger is a dynamic design which can be customized to suit any industrial process depending on the temperature, pressure, type of fluid, phase flow, density, chemical composition, viscosity and many other thermodynamic properties [2, 3]. Due to global energy crisis, an efficient heat recovery or dissipation of heat has become a vital challenge for Scientists and Engineers [4].

INTECH
open science | open minds

© 2017 The Author(s). Licensee InTech. This chapter is distributed under the terms of the Creative Commons Attribution License (http://creativecommons.org/licenses/by/3.0), which permits unrestricted use, distribution, and reproduction in any medium, provided the original work is properly cited.

Heat exchangers are designed to optimize the surface area of the wall between two fluids to maximize the efficiency, while minimizing resistance to fluid flow through the exchangers within constrain of material cost. The performance of heat exchanging surfaces could be enhanced by the addition of corrugations or fins in heat exchanger, which increase surface area and may channel fluid flow or induce turbulence [5]. Efficiency of industrial heat exchangers could be online monitored by tracking the overall heat transfer coefficient based on its temperatures which tends to decline over time due to fouling [6].

Potential damage towards equipment caused by formation of scale can be very costly if processed water is not treated correctly. Chemicals are commonly used to treat the water in the industry. A total of 7.3 billion dollar worth chemicals per year in the U.S. is released into the air, dumped in streams and buried in landfills every year. Forty percent of these chemicals is purchased by industry for control of scale in cooling tower, boiler and other heat transfer equipment. This percentage also represents more than 2 billion dollar of toxic waste which contribute to trillion of gallon contaminated water disposed annually into the earth which belongs to all of us.

Maintenance of fouled tubular heat exchangers can be performed by several methods such as acid cleaning, sandblasting, high-pressure water jet, bullet cleaning or drill rods. In large-scale cooling water systems for heat exchangers, water treatment such as purification, addition of chemicals, catalytic approach, etc., are used to minimize fouling of the heat exchanging equipment [7]. Other water treatment processes are also used in steam systems for power plants to minimize fouling and corrosion of the heat exchanger and other equipment. Most of the chemical and additives used for fouling and corrosion mitigation are hazardous to the environment [8]. So, the days have come to apply chemicals of approaches benign to the environment [9–11].

2. About industrial heat exchanger

An industrial heat exchanger is heat transfer equipment that utilizes a thermal energy exchange process between two or more medium available at different temperature. Industrial heat exchangers are applied in various industrial applications such as power plant generation, petroleum oil and gas industry, chemical processing plant, transportation, alternate fuels, cryogenic, air conditioning and refrigeration, heat recovery and other industries. In addition, heat exchangers are the equipment always closely related to our daily life, for example, evaporators, air preheaters, automobile radiators, condensers and oil coolers. In most heat exchangers, a heat transfer surface separates the fluid which incorporates a wide range of different flow configuration to achieve the desired performance in different applications. Heat exchangers could be classified in many different ways. Generally, industrial heat exchangers have been classified according to construction, transfer processes, degrees of surface compactness, flow arrangements, pass arrangements, phase of the process fluids and heat transfer mechanisms as seen in **Figure 1**.

http://dx.doi.org/10.5772/66274

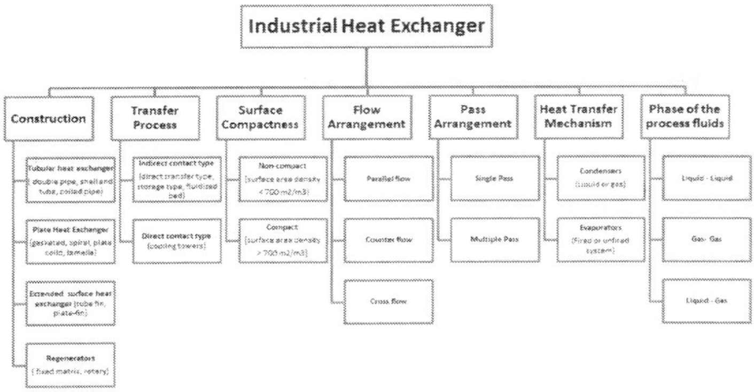

Figure 1. Classification of industrial heat exchanger [12].

3. Basic design concepts for heat exchanger

The design concepts of heat exchanger must meet normal process requirements specified through service conditions for combinations of un-corroded and corroded conditions and the clean and fouled conditions. One of the critical criteria of heat exchanger design is the exchanger must be designed for ease of maintenance, which usually means cleaning or replacement of parts, tubing, fittings, etc. damaged by ageing, vibration, corrosion or erosion throughout the service period.

Hence, a heat exchanger design should be as simple as possible particularly if heavy fouling is expected. By minimize temperature in conjunction with the choice of fluid velocity and reducing the concentration of foulant precursors, will reduce the incidence of potential fouling. Moreover, highest flowing velocity should be allowed under the constraints of pressure drop and erosion from the flow. In addition, material selection within constrained cost retards the build-up of deposits and allows shorter residence time. It should also be compatible in terms of pH, corrosion and not only just heat exchanger, but also in terms of heat equipment and transfer lines of the heat exchanger.

4. Fouling

Fouling is always defined as the formation and accumulation of unwanted materials deposit onto the processing equipment surfaces. These normally very low thermal conductivity materials form an insulation on the surface which can extremely deteriorate the performance of the surface to transfer heat under the temperature difference for which it was designed [13]. On top of this, fouling increases the resistance to fluid flow, resulting in higher pressure drop across the heat exchanger. Many types of fouling can occur on the heat transfer surfaces, for examples, crystallization fouling, particulate fouling, corrosion fouling, chemical reaction fouling, biological fouling and solidification fouling [14]. Fouling can have a very costly effect in the industries which eventually increases fuel usage, interrupts operation, production losses and enhances maintenance costs [15].

The fouling is formed in five stages which can be summarized as initiation of fouling, transport to the surface, attachment to the surface, removal from the surface and ageing at the surface [16]. There are a few parameters influencing the fouling factors, such as pH [9], velocity [17], bulk temperature of fluid [18], temperature of the heat transfer surface, surface structure [19] and roughness [20, 21].

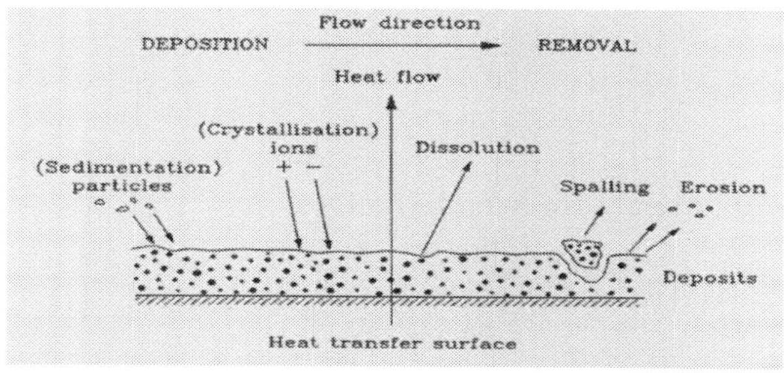

Figure 2. Overall fouling process [22].

The overall fouling process is usually considered to be the net result of two simultaneous sub processes: a deposition process and a removal process as shown in **Figure 2**. As illustrated in

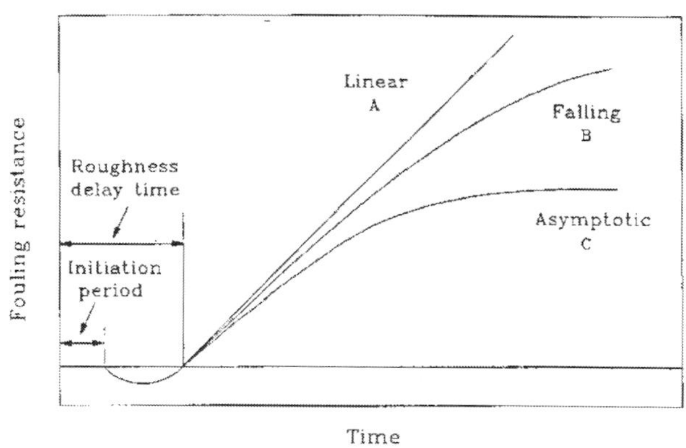

Figure 3. Fouling resistance against time curves [22].

Figure 3, the growth of these deposits causes the heat transfer performance of heat exchanger to decline with time. This problem affects the energy consumption of industrial processes and eventually causes industrial breakdown due to the heat exchanger failure as seen in **Figure 4**.

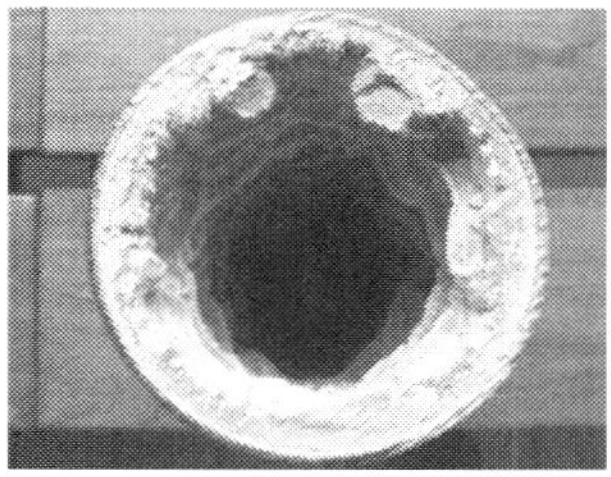

Figure 4. Heavy build-up of deposition on heat exchanger piping [24][23].

5. Corrosion

Environment features such as soil, atmosphere, water or aqueous solutions commonly attack general metal and alloys. The deterioration of these metals is known as corrosion. It is agreeable that corrosion happens due to electrochemical mechanism. Premature failures in various equipment are caused by corrosion in most commercial processes and engineering operations, leading to unwanted issues. This includes pricey breakdown, un-schedule shutdown and increases in maintenance cost.

This downtime worsens in fields such as chemical industries, oil refining, sea and land electric power plant, paper manufacture, air conditioning, refrigerator, food and liquor manufacturing. Hence, general info and mechanism of corrosion will bring great interest to public and

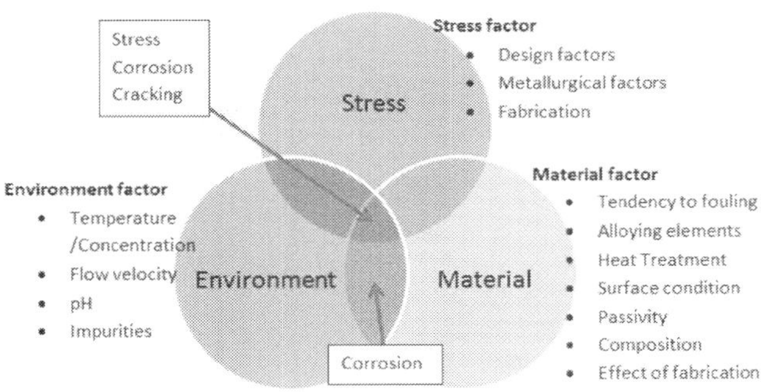

Figure 5. Factor influencing corrosion [25].

industry [24]. The corrosion process is affected by various parameters as show in **Figure 5**. Hence, these criteria should take consideration into the design basics of heat exchangers.

6. Cost imposed due to fouling

Apart from the high cost of heat exchanger fouling, very few work have been reported to accurate determine economic penalties causes by fouling. Therefore, these attribute cost to difference aspect of heat exchanger design and operation. However, reliable knowledge of fouling economics is desirable in order to evaluate the cost efficiency of various mitigation strategies [26, 27]. The total fouling-related costs involve the following:

1. **Capital expenditure**

   Excessive surface area required to overcome the heavy fouling conditions, costs for stronger foundation, provision for extra spaces and increased transportation and installation costs.

2. **Energy costs**

   Costs for extra fuel required if fouling leads to extra fuel burning in heat exchanging equipment to overcome the effect of fouling.

3. **Maintenance costs**

   Costs for removal of fouling deposits, costs for chemicals or other operating costs for antifouling devices.

4. **Cost of production loss**

   Planned or unplanned plant shutdowns due to fouling in heat exchangers can cause large production losses. These losses are often considered to be the main cost of fouling and very difficult to estimate.

5. **Extra environmental management cost**

   Cost for disposing large amount of chemical/additives used for fouling mitigation.

Country	Fouling cost US $ million	GNP (1984) US $ billion	Fouling costs % of GNP
USA	3860–7000 8000–10,000	3634	0.12–0.22 0.28–0.35
Japan	3062	1225	0.25
West Germany	1533	613	0.25
UK	700–930	285	0.20–0.33
Australia	260	173	0.15
New Zealand	35	23	0.15
Total industrial world	26,850	13,429	0.20

Table 1. Estimated fouling costs incurred in some countries (1992 estimation) [28].

http://dx.doi.org/10.5772/66274

Huge fouling costs are reported in different countries. Steinhagen et al. reported about the fouling costs in term of GNPs for some countries as presented in **Table 1**.

7. Current efforts to solve fouling deposition and corrosion problems

A lot of works have been done to reduce fouling formation and control of corrosion. In recent years, numerous methods have been developed to control fouling and corrosion [29]. These methods can be classified as chemical means (inhibitor), mechanical means, changing the phases of the solution, electromagnetic fields, electrostatic fields, acoustic fields, ultraviolet light, radiation or catalytic treatment, surface treatment, green additives, fibre as a suspension, etc. In the past, chromate was a successful chemical agent for corrosion protection and crystal growth control until it was banned. Polyphosphate corrosion inhibitor was introduced for the replacement for chromate-based additives. This inhibitor has a tendency to decompose the foulant in water containing high calcium hardness. Knudsen et al. investigated fouling of high calcium water containing phosphate corrosion inhibitor. Four different copolymers were used to inhibit the deposition of calcium phosphate which includes acrylic acid/maleic anhydride (AA/MA), acrylic acid/hydroxypropyl acrylate (AA/HPA), acrylic acid/sulfonic acid (AA/SA) and sulfonated styrene/maleic anhydride (SS/MA). Studies were carried out by varying the pH, surface temperature and velocities. The investigation reported stated that AA/HPA and (AA/SA) were both very effective in inhibiting deposition of calcium phosphate and corrosion.

On the other hand, catalyst material composed of zinc and tourmaline was studied to mitigate fouling and corrosion. Tijing et al. reported that catalyst material potentially reduces calcium carbonate fouling formation [30]. Teng et al. reported the similar finding of catalyst material on calcium sulphate mitigation [31]. Moreover, Tijing et al. further extended the research by using same catalyst material to mitigate corrosion on carbon steel piping [31].

In the past, most of the methods used, chemical/additives for fouling and corrosion mitigation are hazardous to the environment. So, the days have come to apply green technology method and chemicals approaches benign to the environment [9–11].

8. Fouling mitigation by green technology (catalytic mitigation and green additive)

Physical water treatment (PWT) is a good alternative for safe and efficient nonchemical fouling mitigation method. Examples of PWT include permanent magnets [32], solenoid coil devices [33], green additive [34] and catalytic materials and alloys [35].

To mitigate scaling on heat transfer surfaces, chemical additives are often used, but chemicals are expensive and pose hazard threat to the environment and health. Mitigation of calcium sulphate dehydrates scale formation on heat exchanger surfaces by using natural wood pulp fibre was conducted by Kazi [36] and others in University of Malaya. Experimental work was designed and fabricated to study the use of natural wood pulp fibre as a means of fouling mitigation as seen in **Table 2** and **Figure 6**.

Experimental test rig	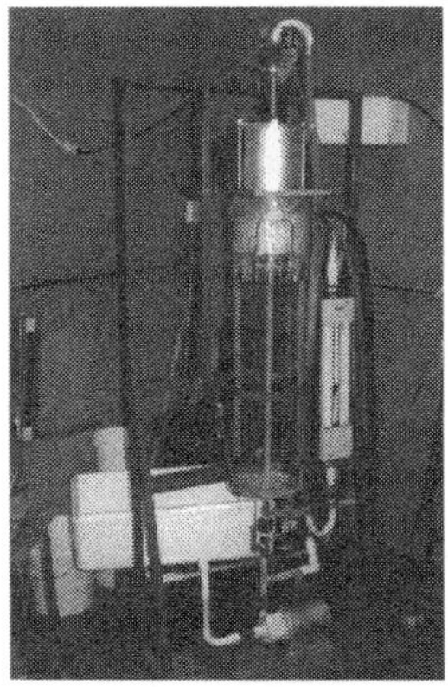
Experimental condition	Fluid velocity = 0.1 m/s Bulk temperature = 30°C ΔTemperature = 15°C $CaSO_4$ = 3.6 g/l C_{fibre} = 0.005–0.1%
Green additive	**Bleached kraft pup fibre** -Freeness–720 CSF -Average length–2.51 mm -Coarseness–0.224
Test specimen materials	SS316, brass, copper, aluminium

Table 2. Experiments set-up for fouling mitigation by incorporating green additives [36][37].

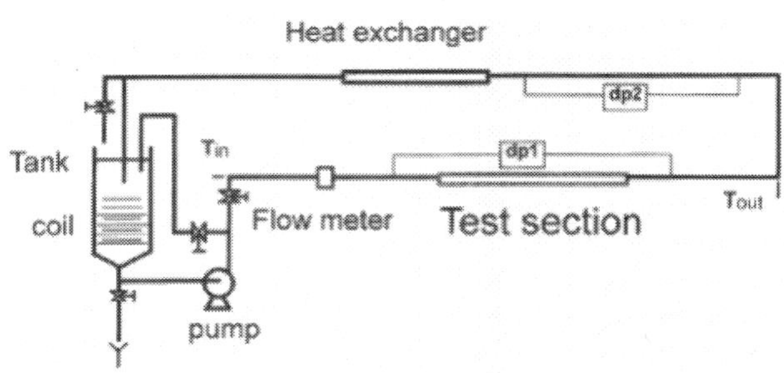

Figure 6. Schematic diagram of experiemental flow loop [37][36].

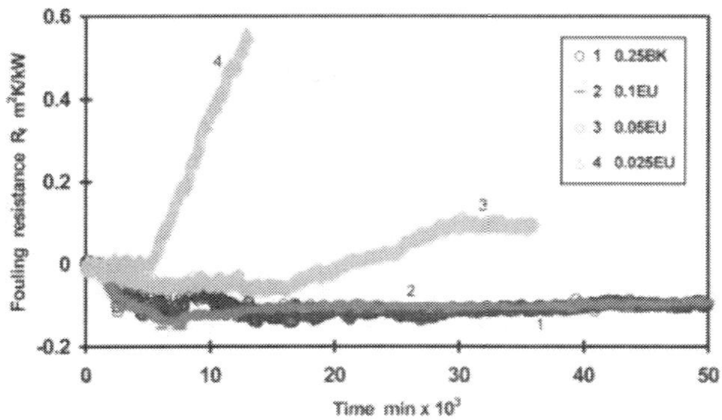

Figure 7. Fouling resistance as a function of time for eucalyptus fibre in the supersaturated calcium sulphate solution [38][37].

Figure 7 depicts the fouling resistance as a function of time for calcium sulphate solution with different fibre concentration of 0.25% (1), 0.15% (2), 0.05% (3) and 0.02% curve (4) in the mineral solution. Results show that the fibres in solution retarded fouling on heated surfaces and the retardation is proportional to the concentration of fibre in the solution. The induction period has also increased.

9. Cleaning of heat exchanger

In order to maintain or restore efficiency of the heat exchanger, it is often necessary to clean the heat exchangers. Methods of cleaning may be classified into two groups: online and offline cleaning [38]. In some applications, the cleaning can be done online to maintain acceptable performance without interruption of operation. In other cases, offline cleaning must be used.

9.1. Online cleaning

Online cleaning generally utilizes a mechanical method designed for only tube side and requires no disassembly. The advantages of online cleaning are the continuity of service of the heat exchanger with the hope that no cleaning-mandated downtime will occur. However, it adds extra cost of a new heat exchanger installation or the large cost of retrofits and there is no assurance that all the tubes would be cleaned sufficiently.

I. Circulation of sponge rubber balls [39]

The technique is capable of preventing the accumulation of particulate matter, biofilm formation and scale and corrosion product deposition. It is only applicable to flow through the inside of tubes.

II. Two phases of the ferrous sulphate treatment

The first phase involves the initial laying down of the protective film. The second phase involves the maintenance of the film, which would be otherwise destroyed by the shear effects of flow.

III. Chlorination used for combat bio-fouling [40]

IV. Scale inhibitors [10, 41, 42]

V. Magnetic devices [10, 43, 44]

VI. Sonic technology [45]

High and low frequency sound emitters (horns) are used to relief fouling problems on heat exchangers. The use of sound is much less effective in sticky and tenacious deposits that are generally associated with slagging.

VII. Online chemical cleaning [46]

Injection of chemical solutions into the process streams for the cleaning purposes.

VIII. The use of radiation [47]

Radiation sterilization of microbial-laden water, the use of ultraviolet light and Gamma rays have been considered for a long time.

9.2. Offline cleaning

An alternative to online cleaning is to stop operation and clean the heat exchanger. Offline cleaning can be classified into offline chemically cleaning or by mechanical means. The cleaning method preferred without the need to dismantle the heat exchangers, but usually it is necessary to have access to the inside surfaces. It would be prudent to consider the installation of a "standby" heat exchanger, thereby providing the opportunity to clean the fouled heat exchanger while at the same time maintain the production.

9.2.1. Offline mechanical cleaning

a. Tube drilling and rodding [28]

Devices may be applied to the rotating shaft including drills, cutting and buffing tools and brushes that may be made from different materials, for example, steels or nylon, brasses depending on the tube material and the nature of the deposit.

b. Cleaning with explosives

Used of controlled explosions, where the energy to remove the deposits, is transmitted by a shock wave in the air adjacent to the surface to be cleaned or by the general vibration of tubes brought about the explosion. It is a relatively new innovation introduced in boiler plant cleaning. It is possible to begin the cleaning process, while the structure is still hot.

c. Thermal shock [48]

Changes in temperature particularly rapid changes, cause cracking of foulant layer with the possibility of flaking. This technique is similar to steam soaking. The water flushing carries away the dislodged material and it is repeated until clean surfaces are obtained.

9.2.2. Offline chemical cleaning

a. d. Inhibitor hydrofluoric, hydrochloric, citric, sulphuric acid or EDTA (chemical cleaning agent) for iron oxides, calcium/magnesium scales (foulant), etc. cleaning [49].

Inhibitor hydrofluoric acid is by far the most effective agent but cannot be used if deposits contain more that 1% w/v calcium.

b. e. Chlorinated or aromatic solvents followed by washing are suitable for heavy organic deposits for example, tars and polymers (foulant) [50].

c. f. Alkaline solutions of potassium permanganate [51] or steam-air decoking [52] are suitable for cleaning carbon (foulant) deposition.

10. Conclusion

Fouling and corrosion are the major unresolved crisis in heat exchanger operation. Though the fouling deposition problems and the impact to the economy are a serious concern, still there is lack of awareness in concerned authorities. In addition, the penalties of corrosion are numerous and varied and the effects of these on the efficient, reliable and safe operation of equipment or structures are often more serious than the simple loss of a mass of metal. Therefore, the present paper will promote concerned organization in different countries, seriousness of this problem and application of possible mitigation approach.

For an industry, the proper cleaning method and control play important role to reduce the production costs. Production cost significantly increases due to chemical usage, maintenance work and downtime loss and water wastage. Consequently, the related authorities need to comprehend the importance of corrosion control, fouling cleaning and enforce a specific standard of cleaning procedure in the industries.

Acknowledgements

The authors gratefully acknowledge to High Impact Research Grant UM.C/625/1/HIR/MOHE/ENG/45, UMRG RP012A-13AET, University Postgraduate Research Fund (PPP) (e.g. PG109-2015A), Liverpool John Moores University United Kingdom and University of Malaya, Malaysia for support to conduct this research work.

Author details

Teng Kah Hou[1,2*], Salimnewaz Kazi[1], Abu Bakar Mahat[1], Chew Bee Teng[1], Ahmed Al-Shamma'a[2*] and Andy Shaw[2]

*Address all correspondence to: alex_teng1989@hotmail.com; k.h.teng@2016.ljmu.ac.uk and salimnewaz@um.edu.my; salimnewaz@yahoo.com

1 Department of Mechanical Engineering, University of Malaya, Kuala Lumpur, Malaysia

2 Department of Built Environment, Liverpool John Moores University, Liverpool, United Kingdom

References

[1] Bott, T.R., *CHAPTER 1 - Introduction, in Fouling of Heat Exchangers.*. 1995, Elsevier Science B.V.: Amsterdam. p. 1–6.

[2] Alahmad, M., *Factors affecting scale formation in sea water environments—an experimental approach*. Chemical Engineering and Technology, 2008. 31(1): pp. 149–156.

[3] Bott, T.R., *Aspects of crystallization fouling*. Experimental Thermal and Fluid Science, 1997. 14(4): pp. 356–360.

[4] Somerscales, E.F.C., *Fouling of Heat Transfer Equipment*. Hemisphere, New York, 1981.

[5] Epstein, N., *Heat Exchanger Theory and Practice*. In Heat exchangers, J. Taborek and G. Hewitt (Eds.), 1983. In Heat Exchanger Theory and Practice: McGraw-Hill.

[6] Panchal, C.B. and J.G. Knudsen, *Mitigation of Water Fouling: Technology Status and Challenges*, In *Advances in Heat Transfer*, T.F.I.Y.I.C. James P. Hartnett and A.G. George, Editors. 1998, Elsevier. pp. 431–474.

[7] Kazi, S.N., et al., *Study of mineral fouling mitigation on heat exchanger surface*. Desalination, 2015. 367: pp. 248–254.

[8] Karagiannis, I.C. and P.G. Soldatos, *Water desalination cost literature: review and assessment*. Desalination, 2008. 223(1–3): pp. 448–456.

[9] MacAdam, J. and S.A. Parsons, *Calcium carbonate scale formation and control*. Re/Views in Environmental Science and Bio/Technology, 2004. 3(2): pp. 159–169.

[10] Macadam, J. and S. Parsons, *Calcium carbonate scale control, effect of material and inhibitors*. Water Science and Technology, 2004. 49(2): pp. 153–159.

[11] Kazi, S.N., G.G. Duffy and X.D. Chen, *Fouling and fouling mitigation on heated metal surfaces*. Desalination, 2012. 288: pp. 126–134.

[12] Thulukkanam, K., *Heat Exchangers*, in *Heat Exchanger Design Handbook, Second Edition*. 2013, CRC Press: United Kingdom. p. 1–38.

[13] Hans, M.-S., *C4 Fouling of Heat Exchanger Surfaces*, in *VDI Heat Atlas*. 2010, Springer Berlin Heidelberg. pp. 79–104.

[14] Bott, T.R., *Fouling in Heat Transfer Equipment*. The Chemical Engineering, 1971: pp. 391–395.

[15] Steinhagen, R., H. Müller-Steinhagen and K. Maani, *Problems and costs due to heat exchanger fouling in New Zealand industries*. Heat Transfer Engineering, 1993. 14(1): pp. 19–30.

[16] Mullin, J.W., *Crystallisation*. 4th ed. 2001, London, UK: Butterworth-Heineman.

[17] Kukulka, D.J. and M. Devgun, *Fluid temperature and velocity effect on fouling*. Applied Thermal Engineering, 2007. 27(16): pp. 2732–2744.

[18] Hoang, T.A., H.M. Ang and A.L. Rohl, *Effects of temperature on the scaling of calcium sulphate in pipes*. Powder Technology, 2007. 179(1): pp. 31–37.

[19] Razmjou, A., J. Mansouri and V. Chen, *The effects of mechanical and chemical modification of TiO_2 nanoparticles on the surface chemistry, structure and fouling performance of PES ultrafiltration membranes*. Journal of Membrane Science, 2011. 378(1): pp. 73–84.

[20] Kazi, S.N., G.G. Duffy and X.D. Chen, *Mineral scale formation and mitigation on metals and a polymeric heat exchanger surface*. Applied Thermal Engineering, 2010. 30(14–15): pp. 2236–2242.

[21] Albert, F., W. Augustin and S. Scholl, *Roughness and constriction effects on heat transfer in crystallization fouling*. Chemical Engineering Science, 2011. 66(3): pp. 499–509.

[22] Mostafa M. Awad (2011). Fouling of Heat Transfer Surfaces, Heat Transfer – Theoretical Analysis, Experimental Investigations and Industrial Systems, Prof. Aziz Belmiloudi (Ed.), InTech, DOI: 10.5772/13696. Available from: http://www.intechopen.com/books/heat-transfer-theoretical-analysis-experimental-investigations-and-industrial-systems/fouling-of-heat-transfer-surfaces

[23] S. N. Kazi (2012). Fouling and Fouling Mitigation on Heat Exchanger Surfaces, Heat Exchangers – Basics Design Applications, Dr. Jovan Mitrovic (Ed.), InTech, DOI: 10.5772/32990. Available from: http://www.intechopen.com/books/heat-exchangers-basics-design-application/heat-exchanger-fouling-and-its-mitigation

[24] M. G. Fontana, Corrosion Engineering, McGraw-Hill International Edition, 3rd ed., pp. 23–27 and pp. 499–503.

[25] Thulukkanam, K., Corrosion, in *Heat Exchanger Design Handbook, Second Edition*. 2013, CRC Press: United Kingdom. p. 665–748

[26] Teng, K.H., et al., *Retardation of heat exchanger surfaces mineral fouling by water-based diethylenetriamine pentaacetate-treated CNT nanofluids*. Applied Thermal Engineering, 2017. 110: pp. 495–503.

[27] Teng, K.H., et al., *Fouling mitigation on heat exchanger surfaces by EDTA-treated MWCNT-based water nanofluids*. Journal of the Taiwan Institute of Chemical Engineers, 2016. 60: pp. 445–452.

[28] Müller-Steinhagen, H., M. Malayeri and A. Watkinson, *Heat exchanger fouling: mitigation and cleaning strategies*. Heat Transfer Engineering, 2011. 32(3–4): pp. 189–196.

[29] Alahmad, M., *Experimental study of scale formation in sea water environment*. Journal of King Saud University, 2004. 17(1): pp. 73–88.

[30] Tijing, L.D., et al., *Mitigation of scaling in heat exchangers by physical water treatment using zinc and tourmaline*. Applied Thermal Engineering, 2011. 31(11): pp. 2025–2031.

[31] Teng, K., et al., *Mitigation of heat exchanger fouling in industry using catalytic materials*. Desalination and Water Treatment, 2016. 57(1): pp. 22–27.

[32] Lee, G.J., et al., *Use of catalytic materials for the mitigation of mineral fouling*. International Communications in Heat and Mass Transfer, 2006. 33(1): pp. 14–23.

[33] Tijing, L.D., et al., *Physical Water Treatment Using Oscillating Electric Fields to Mitigate Scaling in Heat Exchangers*, in *Frontiers and Progress in Multiphase Flow I*. 2014, Springer. p. 105–155.

[34] Hasson, D., H. Shemer and A. Sher, *State of the art offriendly "green" scale control inhibitors: a review article*. Industrial and Engineering Chemistry Research, 2011. 50(12): pp. 7601–7607.

[35] Brinis, H. and M.E.H. Samar, *A method for inhibiting scale formation and corrosion in a cooling water system*. Desalination and Water Treatment, 2013. 52(13–15): pp. 2609–2619.

[36] Kazi, S.N., G.G. Duffy and X.D. Chen, *Fouling mitigation of heat exchangers with natural fibres*. Applied Thermal Engineering, 2013. 50(1): pp. 1142–1148.

[37] Kazi, S.N., G.G. Duffy and X.D. Chen, *Fiber-modified scaling in heat transfer fouling mitigation*. Chemical Engineering Communications, 2002. 189(6): pp. 742–758.

[38] MÜLler-Steinhagen, H., M.R. Malayeri and A.P. Watkinson, *Fouling of heat exchangers-new approaches to solve an old problem*. Heat Transfer Engineering, 2005. 26(1): pp. 1–4.

[39] H. Müller-Steinhagen, M. R. Malayeri, A. P. Watkinson. (2011) Heat Exchanger Fouling: Mitigation and Cleaning Strategies. Heat Transfer Engineering 32:3–4, pages 189–196.

[40] Rajagopal, S., et al., *How effective is intermittent chlorination to control adult mussel fouling in cooling water systems?* Water Research, 2003. 37(2): pp. 329–338.

[41] Choi, D.-J., S.-J. You and J.-G. Kim, *Development of an environmentally safe corrosion, scale and microorganism inhibitor for open recirculating cooling systems*. Materials Science and Engineering: A, 2002. 335(1–2): pp. 228–235.

[42] Lin, Y.-P. and P.C. Singer, *Inhibition of calcite crystal growth by polyphosphates*. Water Research, 2005. 39(19): pp. 4835–4843.

[43] Parson, S.A., et al., *Magnetically augmented water treatment*. Process Safety and Environmental Protection, 1997. Transaction of the Institution of Chemical Engineering 75(Part B): pp. 98–104.

[44] Parsons, S.A., et al., *Magnetic treatment of calcium carbonate scale—effect of pH control*. Water Research, 1997. 31(2): pp. 339–342.

[45] Lin, S. and X. Chen, *A laboratory investigation of milk fouling under the influence of ultrasound*. Food and Bioproducts Processing, 2007. 85(1): pp. 57–62.

[46] Chen, X.D., et al., *On-line fouling/cleaning detection by measuring electric resistance—equipment development and application to milk fouling detection and chemical cleaning monitoring*. Journal of Food Engineering, 2004. 61(2): pp. 181–189.

[47] Park, J.-S., et al., *Fouling mitigation of anion exchange membrane by zeta potential control*. Journal of Colloid and Interface Science, 2003. 259(2): pp. 293–300.

[48] Wilson, D., *Challenges in cleaning: recent developments and future prospects*. Heat Transfer Engineering, 2005. 26(1): pp. 51–59.

[49] Madaeni, S. and S. Samieirad, *Chemical cleaning of reverse osmosis membrane fouled by wastewater*. Desalination, 2010. 257(1): pp. 80–86.

[50] Arhancet, G.B., *Compositions and methods for inhibiting fouling of vinyl monomers*. 1999, Google Patents.

[51] Spicher, R.G. and R.T. Skrinde, *Potassium permanganate oxidation of organic contaminants in water supplies*. Journal American Water Works Association, 1963: pp. 1174–1194.

[52] Clavenna, L., et al., *Method for reducing fouling in a refinery*. 2006, Google Patents.